Analyzing and Modeling Interdisciplinary Product Development

Frank Neumann

Analyzing and Modeling Interdisciplinary Product Development

A Framework for the Analysis of Knowledge Characteristics and Design Support

With a Preface by
Univ.-Prof. Dr.-Ing. Dr. h.c. Sándor Vajna

 Springer Vieweg

Frank Neumann
Berlin, Germany

Dissertation, Otto-von-Guericke-University Magdeburg, 2014

ISBN 978-3-658-11091-8   ISBN 978-3-658-11092-5 (eBook)
DOI 10.1007/978-3-658-11092-5

Library of Congress Control Number: 2015951982

Springer Vieweg
© Springer Fachmedien Wiesbaden 2015

Printed on acid-free paper

Springer Vieweg is a brand of Springer Fachmedien Wiesbaden
Springer Fachmedien Wiesbaden is part of Springer Science+Business Media
(www.springer.com)

Preface

During product development, in particular when it no longer focuses on a single area, but, like Mechatronics, takes place in an interdisciplinary environment, on the one hand the product developer needs knowledge of the area of application and product behavior in sufficient quantity and appropriate forms for decision-making, that should be made available in a timely manner, especially since he/she creates comprehensive knowledge on his/her own. It should not only be "producible" and be "manageable" at reasonable cost, but it must be possible to also be incorporated with a similar effort to the existing knowledge landscape. On the other hand, he/she needs such frameworks for his/her interdisciplinary development work that not only aim at creation and acquisition of knowledge, but also provide support for the actual development activities. This is point at which the present dissertation of Mr. Neumann starts.

At first, the reader will find a compact historical and technical introduction to Mechatronics. Thereby, mechatronic product development (MPD) is described as interdisciplinary collaboration between the involved domains, characterized by a high degree of parallelization of activities and the extensive use of models (and related interfaces). In the following, knowledge is considered a *polymorphic* concept, i.e. a concept that cannot be defined by a classical definition comprehensively describing their specific traits. Therefore, the definition of knowledge requires a specific context, which at the same time constitutes the area of validity for this definition.

For the context of interdisciplinary product development, the author emphasizes the dichotomy of tacit knowledge (subjective, experienced) and explicit knowledge (can be objectified and externalized). The interdisciplinary development work is perceived as interplay of individual and collective processes. The author developed a descriptive model of knowledge creation for the associated continuous conversions between tacit and explicit knowledge-as well as a suitable analysis and modeling method. The application of this method to MPD allows the build-up of a library of mechatronic process elements, contributing significantly to an improved ability to model these activities as well as to greater transparency of the activities. By combining these analysis results with the vast potentials of semantic technologies, the author developed a design support system in MPD. Thereby, the knowledge embedded in the utilized models may be captured, extracted, and stored by means of common vocabularies.

Although the author is not a native English speaker, he has written his thesis in fluent and well readable English. The reader may thereby benefit not only from the presentation of the scientific content, but also from the appealing English of this dissertation.

Magdeburg   Univ.-Prof. Dr.-Ing. Dr. h.c. Sándor Vajna

Acknowledgments

My special appreciation goes to my supervisor Prof. Dr.-Ing. Dr. h.c. Sándor Vajna, the holder of chair of Information Technologies in Mechanical Engineering, for the confidence shown in me, the expert suggestions and the various discussions of the topics in the thesis. This granted a freedom that was very important to me during the creation of the present thesis, especially is it allowed me to explore previously unknown fields of knowledge. Likewise, I learned to appreciate the openness to new ideas, which do not always yield the desired result, but often lead to new insights.

I warmly thank Prof. Dipl.-Ing. Dr. Klaus Zeman, the head of the Institute of Mechatronic Design and Production at the Johannes Kepler University of Linz, for his interest in the thesis, for agreeing to act as second examiner and the resulting stimulating discussions.

I would also like to thank the staff of chair of Information Technologies in Mechanical Engineering, in particular Ms. Ina Meseberg, for the organizational support.

I would also like to thank Prof. Dr.-Ing. habil. Dr. h.c. Dr. h.c. Günter Höhne, the former head of the Engineering Design Group at the Technical University of Ilmenau, for the interest concerning questions of design methodology raised and promoted during my studies of precision mechanics/mechanical engineering.

I want to thank Dr. Kilian Gericke for enriching discussions on disciplinary interaction models and M.Sc. Michael Henrichs for the interesting discussions on semantic technologies and for proof reading the respective chapter. My thanks also go to Dr. John Wilkes, who took care for the correct wording, orthography, and grammar as native English speaker.

I would like to sincerely thank my family and friends, especially my life partner Uta Syrbe, for their understanding, wide range of support, and motivation. I know that the time has sometimes not been sufficient to cover all interests alike.

I am indebted to the three managing directors of my employer PACE Aerospace Engineering & Information Technology GmbH, for their benevolent support and helpful arrangements of my working time.

Berlin   Frank Neumann

Abstract

Based on the understanding of Mechatronic Product Development (MPD) as an interdisciplinary activity, both individual and collective activities in product development have to be considered when conceiving design support for MPD. Consequently, the thesis focuses on a first step of establishing a theoretical basis that allows a description of the interplay between individual and collective processes in product development and the sources of knowledge applied within these activities. For this purpose, the integrated descriptive model of knowledge creation is introduced as the first constituent of the overall research framework. It employs a model of cognitive activities in design for explaining the patterns of knowledge application and creation within individuals, whereas it adopts a model of organizational knowledge creation to describe collective processes. For the integration of these approaches, the integrated descriptive model of knowledge creation in interdisciplinary product development follows an approach that merges the two models based on their common conceptual elements. As a second part of the research framework, an analysis and modeling method is proposed that captures the various knowledge conversion activities described by the integrated descriptive model of knowledge creation.

Subsequently, the research framework is applied to the analysis of knowledge characteristics of MPD, whereby the development process is represented by common process elements compiled from various procedure models of MPD. For the first steps of this analysis process, the process and activity views are established for a set of representative process elements. Based on these views, the knowledge characteristics of the process elements are extracted and documented.

Next, the results gained from the previous research steps are used within a design support system that aims at improving the creation, storage, distribution, and context-sensitive provisioning of information and knowledge throughout MPD. In particular, the previously determined characteristics of common process elements for MPD are included within the design support system as a ready-to-use library of common process elements that provides the basis for context-dependent provisioning of information and knowledge objects. In addition, common vocabularies are applied for the semantic enrichment of models used in MPD. They allow extraction of information and knowledge embedded in these models and make them accessible to applications sharing the same vocabularies.

Finally, a system architecture for the design support system is developed that conforms to Linked Data principles (cf. section 8.2.1) and the architectural style called Representational State Transfer (REST). It aims at federating the information and knowledge resources contained in the models published in the course of the development activities in MPD.

Table of Contents

List of Figures

List of Tables

List of Definitions

List of Abbreviations

AI	Artificial Intelligence
AMS	Autonomous Mechatronic System
ASIC	Application Specific Integrated Circuit
BEM	Boundary Element Method
B-Rep	Boundary Representation
BOM	Bill of Material
CACD	Computer-Aided Control Design
CAD	Computer-Aided Design
CAE	Computer-Aided Engineering
CAx	An umbrella term for Computer-Aided Systems, including CAD, CAE, EDA, FEA, etc.
CASE	Computer-Aided Software Engineering
CFD	Computational Fluid Dynamics
DESS	Differential Equation System Specifications
DEVS	Discrete Event System Specifications
D-I-K	Data, Information, and Knowledge
DMU	Digital Mock-Up
DRM	Design Research Methodology
DSL	Domain-specific Language
DTSS	Discrete Time System Specifications
EDA	Electronic Design Automation
E/E	Electrics/Electronics
EEPROM	Electrically Erasable Programmable Read-Only Memory
EPC	Event Driven Process Chain
FEA	Finite Element Analysis
HiL	Hardware-in-the-loop
HTML	Hypertext Markup Language
IA	Intelligent Agent
ID	Identifier
IS	Information Science
IT	Information Technology
JAIST	Japan Advanced Institute of Science and Technology
KBE	Knowledge-Based Engineering
KFAM	Knowledge Flow Analysis and Modeling
KM	Knowledge Management
KMDL	Knowledge Modeling Description Language

KR	Knowledge Representation
KS	Knowledge Science
LIN	Local Interconnect Network
MBD	Model-based Development
MBS	Multibody Simulation
MFM	Mechatronic Functional Module
MPD	Mechatronic Product Development
n/a	Not applicable
NMS	Networked Mechatronic System
OECD	Organization for Economic Co-operation and Development
OMG	Object Management Group
OWL	Web Ontology Language
PDM	Product Data Management
PLC	Programmable Logic Controller
PLM	Product Lifecycle Management
RDF	Resource Description Framework
RDF-S	RDF-Schema
REST	Representational State Transfer
ROA	Resource-oriented Architecture
SECI	Socialization, Externalization, Combination, Internalization: The four modes of organizational knowledge creation
SiL	Software-in-the-loop
SOA	Service-oriented Architecture
SOAP	Simple Object Access Protocol
SPARQL	RDF Query Language
STEP	Standard for Exchange of Product Model Data
SysML	Systems Modeling Language
TASS	The Autonomous Set of Systems
UML	Unified Modeling Language
URI	Uniform Resource Identifier
URL	Uniform Resource Locator
URN	Uniform Resource Name
VDA	Verband der Automobilindustrie: German Association of the Automotive Industry
VDI	Verein Deutscher Ingenieure: Association of German Engineers
VDMA	Verband Deutscher Maschinen- und Anlagenbau: German Engineering Federation
W3C	World Wide Web Consortium
XML	Extensible Markup Language

1 Introduction and Motivation

The complaint about the sharpness of competition is in reality mostly only a complaint about the lack of ideas.[1]

Walter Rathenau, 1918

Walther Rathenau, an influential German industrialist, politician, and writer at the beginning of the 20th century, reveals with crisp irony the close ties between innovative ideas and competiveness. Just as in Rathenau's time, 100 years ago, companies today have to concentrate on maintaining and preferably improving their competitive position under constantly changing market conditions. In this spirit, each company must develop a well-adapted competitive strategy when entering a new market, and once present in this market, further adapt this strategy to the conditions and trends of the particular industry and the characteristics of the market (Porter 1998).

Sometimes it may be sufficient to adapt the company's operations solely in a quantitative direction, e.g. to adapt the overall cost structure by managerial action, or to massively extend the production capacity. Under today's economic and technical conditions, however, companies are rather seeking a competitive advantage through qualitatively new approaches and technologies, e.g. through *"innovation in its broadest sense"* (Porter 1990). Porter states that innovation may manifest itself under different forms: *"a new product design, a new production process, a new marketing approach, or a new way of conducting training"*.

Vajna et al. (2009) elaborate on the contemporary economic conditions under which firms in industrial countries operate and compete, and likewise emphasize the decisive role of innovation:

- Their relatively high costs lead to high product prices that can be realized on the market, only if the products differentiate themselves sufficiently from the competition. Superior functional abilities and higher reliability are examples of attributes that set a product apart from its competitors. This leads to the need to achieve and maintain an advantage by innovation.
- Such high product prices can be fetched on the market only for a relatively short period, as long as the product faces no relevant competition. This forces companies to develop competitive products in ever-shorter cycles, i.e. to reduce their *time to market*.
- In order to benefit from a maximum period of profitability, even for innovative products a strict cost control must be executed.
- Less innovative products or products in a saturated market may only be sold on the basis of their low price, and require therefore low product costs.

---

[1] In original: „Die Klage über die Schärfe des Wettbewerbes ist in Wirklichkeit meist nur eine Klage über den Mangel an Einfällen."

Under the described economic conditions, the development, manufacturing, and successful market introduction of new or enhanced products is challenging. Among other things, it requires the discovery of new sources of innovation and the ability to unlock their inherent potential on a continuous base. In the area of product development, a study of the *Berliner Kreis*[2] on *"Innovation Potentials in Product Development"*[3] presents 16 methods, processes, and systems providing a potential for innovation (Krause, Franke, et al. 2007). Resulting from a ranking of these topics according to their importance for the industry, the study emphasizes three areas of innovation under the special focus of the industry: Mechatronics belongs to this outstanding triple.

The term *Mechatronics* stands for the technology and the products emerging from the ongoing transformation of formerly mechanical products through the addition of electrical components, electronics, and information processing. As the first step of this transformation process, electronics took over certain functions previously provided by mechanical components[4]. In the next step, mechatronic systems were able to supply extended and new features (Isermann 2008). Mechatronics benefits from synergies[5] of the interaction of the involved engineering fields – mechanical engineering, electrical engineering, and information technology (VDI 2004).

More precisely, the highly beneficial potential attributed to Mechatronics arises from two main sources (Zohm 2003; VDI 2004):

(a) The innovation potential of each technology contributing to the field of Mechatronics.
(b) The beneficial potentials obtained from the functional and spatial integration of these technologies.

The speed of innovation in the fields of electronics and software is higher than in the domain[6] of mechanics. Mechatronic systems benefit therefore largely from the aforementioned innovation dynamics in electronics and software through the integration of such components within a mechanical basis (Eversheim, Niemeyer, et al. 2000; Zohm 2003; VDI 2004). The tight combination of the technologies participating in mechatronic products leads to a higher density of functions per volume and to completely new features (Zohm 2003; VDI 2004). At the economic side, the constantly falling prices of microelectronic components offer an strategic

---

[2] The *Berliner Kreis* was an association of 30 professors that acted as a competence network to enhance product innovation in mechanical engineering and associated industries. In 2011, it acted as a founding member for the *Wissenschaftliche Gesellschaft für Produktentwicklung* (WiGeP).

[3] In original: „Innovationspotenziale in der Produktentwicklung"

[4] The section 2.2 provides an in-depth description of the evolutionary changes from purely mechanical, over electro-mechanical, to mechatronic products.

[5] Synergy: Interaction of two or more elements creating an effect that is larger than the summation of their single contributions (Oxford Dictionaries 2010).

[6] In the scope of the present thesis, the terms *„discipline"*, *„domain"* and *"field"* will be used as synonyms. They designate a scientific and professional branch of knowledge.

cost saving potential for producers and integrators of mechatronic devices (Zohm 2003; VDI 2004).

What are the causes for the remarkable interest in Mechatronics displayed by parts of German industry? The OECD's economic survey on Germany (2010) elaborates on the specifics of the German industry: It is the world's largest producer of so-called *"medium-high-tech"*[7] manufacturing goods, with machinery, equipment, and transport vehicles accounting for the largest part, as indicated by Figure 1.1.

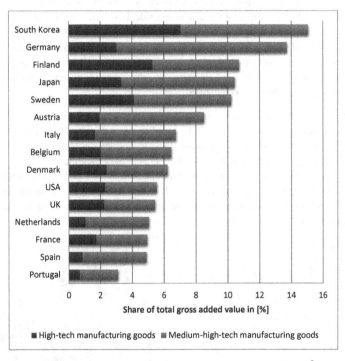

Figure 1.1: Contribution to the total gross added value by categories of manufacturer goods[7], according to (OECD 2010)

In such industries, the relative expenditure for research and development is significantly lower than in *"high-tech"*[7] industries, as e.g. in the aerospace and pharmaceutical industry. In spite of the merely average spending on research and development, innovations are vital for the main German industries of machinery, equipment, and transport vehicles. Here, Mechatronics offers a smart solution for this obvious dilemma - innovation in the form of mechatronic systems may be purchased from system and component suppliers, and integrated into

---

[7] OECD industry classification based on the intensity of research and development (R&D):
High-Technology Industries: R&D intensity > 5%
Medium-High-Technology Industries: R&D intensity between 3% and 5%
(OECD 2007)

the final product, e.g. a car, by the final manufacturer causing only incremental changes to the product's basic structure and functions (Zohm 2003).

These sharp differences in the distribution of challenges and benefits between suppliers and final producers within the automotive industry induced by the change toward Mechatronics characterizes Zohm (2003) as *"dichotomy of change"*:

- The final manufacturer benefits from the technical and economical potential of Mechatronics through the integration of mechatronic systems purchased from suppliers. The changes induced at the final manufacturer's side, however, remain rather incremental, and may be characterized as *"continuous-evolutionary"*(Zohm 2003).
- The changed technological possibilities and the extended customer demands toward functionality, comfort, and security confront suppliers with a radically changing situation. Hence, Mechatronics induces at the supplier's side *"discontinuous-revolutionary"* changes (Zohm 2003).

The German automotive industry, represented by the VDA as its interest group, is convinced that 90% of future innovations in cars will be based on Mechatronics (VDA 2005). Likewise within the German machinery and equipment industry, mechatronic components and systems play an increasingly important role (Synek 2003). In both industries, electronics and software contribute ever more to the implemented product functions and the added value, whereas the importance of mechanics is declining (Dais 2004; Kühnl 2007). Figure 1.2 depicts these past and future trends of the disciplinary mix for the machinery industry according to ITQ GmbH (2010) based on data from the VDMA.

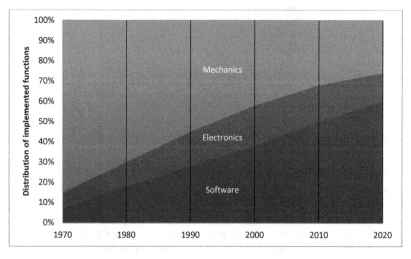

Figure 1.2:   Past and future changes of the disciplinary development mix within the machinery industry, according to (ITQ GmbH 2010)

Although the aforementioned *"discontinuous-revolutionary"* changes for manufacturers of mechatronic systems originate from the largely changed technological basis, this transformation reaches beyond the technological level and induces changes within the product development processes (Zohm 2003). At the time when mechanics still contributed most of the functions and of the value to the final product, a sequential product development process was established: First, a mechanical concept for the product was established, then requirements within electronics and software were taken into account (Eversheim, Niemeyer, et al. 2000; VDI 2004; Angerbauer, Buck, et al. 2010). Within mechatronic product development (MPD), the roughly equal importance of the three engineering disciplines (cf. Figure 1.2), requires a shift toward a multidisciplinary development process (Neumann 2012). Furthermore, the demand for systems optimized over the scope of all involved disciplines dictates tight cooperation, integration, and synchronization of all involved engineering fields (Neumann 2012). In short, the development of mechatronic systems requires an interdisciplinary process (Eversheim, Niemeyer, et al. 2000; VDI 2004).

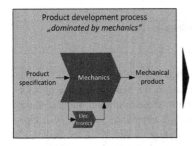 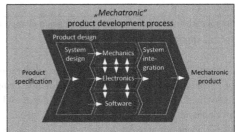

Figure 1.3:   From the mechanical to the mechatronic product development process, according to (Eversheim, Niemeyer et al. 2000)

As depicted on the right side of Figure 1.3, such a mechatronic product development process consists basically of three major phases (Eversheim, Niemeyer, et al. 2000):

(a) The system design phase aims at establishing a solution concept.

(b) The domain-specific design phase further elaborates the elements comprising the solution concept within the context of their respective domain. The intermediate results of each discipline have to be coordinated with the other domains at so called *"synchronization points"*.

(c) During the system integration phase, the discipline-specific results are integrated within the context of the overall system.

Due to the intensified cross-functional interactions, the mechatronic development process implies an increased exchange of information and extended knowledge flows between the involved disciplines (Neumann 2012). Such an interdisciplinary collaboration imposes new requirements on the competences of the experts involved (VDI 2004). For instance, the participants in MPD need to develop knowledge and understanding for the concepts and abilities of the other disciplines (VDI 2004). The Aberdeen Group (2008) reviewed the MPD practices of approx. 140 companies. In the resulting study, problems in the area of system-level knowledge and cross-functional knowledge were identified as the major challenge for successful MPD (cf. Table 1.1).

Table 1.1:   Results of a survey on challenges of mechatronic product development, according to (Boucher and
Houlihan 2008)

Challenges	Response
Difficulty finding/hiring experienced system engineers / lack of cross-functional knowledge	50%
Early identification of system level problems	45%
Ensuring all design requirements are met in the final system	40%
Difficulty predicting / modeling system product behavior until physical prototypes exist	32%
Difficulty implementing an integrated product development solution for all disciplines involved in mechatronic product development	28%
Inability to understand the impact a design change will have across disciplines	18%

Likewise, the *"aquimo"* guideline for establishing tools and qualifications for MPD within the German machinery and equipment industry emphasizes the high importance of information and knowledge level issues for a successful transformation toward interdisciplinary development processes (Angerbauer, Buck, et al. 2010). Furthermore, this guideline highlights the specific interest for strategic knowledge management within MPD due to the high demand for the accumulation and sharing of know-how (Angerbauer, Buck, et al. 2010).

1.1 Problem Analysis

As previously mentioned, the move from a conventional to a mechatronic product development process is characterized as a *"discontinuous-revolutionary"* change for manufacturers of mechanical components developed further toward mechatronic systems (Zohm 2003). For such a transformation, numerous challenges lie ahead that have to be coped with until Mechatronics' innovation potential may be fully realized (Neumann 2012).

An analysis of the relevant literature on MPD, e.g. academic textbooks, engineering guidelines, research reports, and dissertations published since 1999, provides the input for the collection, structuring, and systematization of problems encountered in MPD. The matches received as results from a first round of the literature search on challenges in MPD, underwent a subsequent analysis, keeping only such references that elaborate with adequate breadth on problems in MPD. Eight publications met these criteria and served therefore as input set for the final step of structuring and systematization according to areas of criteria described by the research framework for Mechatronics proposed by Eversheim et al. (2000). This framework aims at linking the areas of (a) products and technologies, (b) product development processes and (c) structuring of organizations, IT, knowledge, and qualification within interdisciplinary research on MPD covering social, economic, and technical aspects (Neumann 2012). Eppinger (2002) proposed a closely related but comparatively limited model for complexity studies consisting of the domains (a) product, (b) process and (c) organization, omitting aspects like IT, knowledge and qualification required for the present thesis.

1.1.1 Product and Technology Related Challenges

Table 1.2 depicts the challenges resulting from the integration of electronics and information processing into formerly mechanical products. Most of the analyzed references acknowledge the complexity (ID 1.1 in Table 1.2) resulting from the integration of these technologies as the most prominent challenge. The complexity of mechatronic products arises from the high number of cross-disciplinary relations, i.e. interactions of elements in different engineering domains (VDI 2004; Krause, Franke, et al. 2007). El-khoury (2006) mentions the increase of software-based functions as another source of product complexity. Some authors explicitly mention technological heterogeneity (ID 1.2) as another source of product complexity (Eversheim, Niemeyer, et al. 2000; VDI 2004; Bellalouna 2009). In conclusion, El-khoury (2006) describes the challenge resulting from the high product complexity as *"to handle systems of 'organised complexity' - systems that are too complex for analysis and too organised for statistics"*.

On the one hand, reliability (ID 1.3) is mentioned in a few sources as a challenge for mechatronic systems (Krause, Franke, et al. 2007; Bellalouna 2009) caused by the mounting pressure on cost and the reduced development cycles. For example, Bellalouna (2009) elaborates observations of increasing reliability problems for mechatronic systems in the automotive industry and identifies problems arising from failures in product development as the cause for 80% of these defects. On the other hand, mechatronic products bear the potential for higher reliability due to the reduced number of interfaces and connectors (VDI 2004).

Table 1.2: Product and technology related challenges of mechatronic products

Sources	Text-books	Engineering Guidelines		Research Reports		Dissertations		
	Krause et al. (2007), Innovationspotenziale	VDI 2206 (2004)	Angerbauer et al. (2010), aquimo	Aberdeen Group (2008), System Design	Eversheim et al. (2000), Chancen	Bellalouna (2009), Integrationsplattform	El-khoury (2006), Model Management	Kümmel (1999), Integration
ID	**Challenges**							
1.1	Complexity due to the high number of cross-disciplinary relations	●	●	●	●		●	●	
1.2	Technological heterogeneity contributes to product complexity and requires cross-disciplinary coordination		●			●	●		
1.3	Potential for higher reliability enabled by reduced number of interfaces but difficult to realize due to reduced development time and increased cost pressure	●	●				●		

1.1.2 Process Related Challenges

The following three sections focus on the challenges induced by the transformation of the product development process toward an interdisciplinary product development process. In particular, the references in the literature will be analyzed for problems arising from (a) the multidisciplinary nature of MPD, (b) the needs for virtual prototyping and model-based development, and (c) the needs for holistic specification of mechatronic systems and end-to-end tracking of requirements during MPD.

1.1.2.1 Problems Related to Multidisciplinarity

Table 1.3 gives an overview on the problems induced by the multidisciplinary nature of MPD. Almost all of the analyzed references confirm that the transition to an interdisciplinary and integrated product development process is challenging due to complex cross-disciplinary coordination and collaboration tasks (ID 2.1).

Table 1.3: Challenges arising from the multidisciplinary nature of the mechatronic development process

Sources		Text-books	Engineering Guidelines		Research Reports		Dissertations		
		Krause et al. (2007), Innovationspotenziale	VDI 2206 (2004)	Angerbauer et al. (2010), aquimo	Aberdeen Group (2008), System Design	Eversheim et al. (2000), Chancen	Bellalouna (2009), Integrationsplattform	El-khoury (2006), Model Management	Kümmel (1999), Integration
ID	**Challenges**								
2.1	Move to interdisciplinary and integrated development process difficult due to the complex cross-disciplinary coordination and collaboration tasks	●	●	●	●		●	●	●
2.2	Interdisciplinary collaboration hindered by differences in terminology, methods and procedures between disciplines	●				●			●
2.3	Holistic exploitation of product not possible in early design phases				●	●			
2.4	Difficult to understand what cross-disciplinary impact a design change will have				●		●		
2.5	Domain-spanning integration-, configuration-, change and version management only poorly supported						●		

Each of the disciplines involved in MPD possesses its own terminology, methodologies, methods, and tools (Krause, Franke, et al. 2007). These disciplinary differences impede the collaboration between the different domains (ID 2.2) and typically lead to *"inconsistent partial developments and unaccounted interactions"* (Krause, Franke, et al. 2007). As a correction for these problems, Eversheim et al. (2000) propose a continuous coordination of disciplinary results at so called *"synchronization points"*, based on well-defined interfaces between the different disciplinary department.

Another problem pointed out by two of the analyzed sources is the missing ability for holistic exploitation of the product as integrated mechatronic system during early design phases (ID 2.3). To address this issue, a system design phase has been proposed that aims at handling system level problems in early design stages (Eversheim, Niemeyer, et al. 2000; VDI 2004; Boucher and Houlihan 2008). During this system design phase, the solution concept of the product will be conceived based on the overall functionality divided up into sub-functions.

Two references mention the difficulty of understanding the impact of a design change originated by one discipline on the fulfillment of product requirements and on the other involved disciplines (ID 2.4) (Boucher and Houlihan 2008; Bellalouna 2009). In order to be able to fully understand the consequences of design changes, a consistent mapping between design requirements and components of the mechatronic system is required that enables the designer to (a) identify the design requirements affected by the changed component and (b) detect all items in the different disciplines involved in the fulfillment of these particular requirements.

Another problem pointed out by Bellalouna (2009) is the difficulty in establishing domain-spanning processes for e.g. integration, configuration and change management (ID 2.5). Although each of the disciplines involved in MPD typically practices such coordination and synchronization methods, these disciplinary processes need to be complemented and coordinated by domain-spanning counterparts acting on the level of the mechatronic product. For this purpose, Bellalouna (2009) proposes a function-oriented synchronization approach for the discipline-specific processes of version, approval, change and configuration management. The proposed approach acts at the interdisciplinary level of product functions and employs therefore a function-oriented point of view to determine all functionally related components and systems (independently of their disciplinary adherence) when making decision for the versioning, approval, change or configuration management in the mechatronic context.

1.1.2.2 Challenges within Virtual Prototyping and Model-based Development

Table 1.4 lists the challenges in the areas of virtual prototyping[8] and model-based development[9] found in the analyzed research literature.

---

[8] Virtual Prototyping: *"By virtual prototyping, we refer to the process of simulating the user, the product, and their combined (physical) interaction in software through the different stages of product design, and the quantitative performance analysis of the product."* (Song, Krovi, et al. 1999; cited in Wang 2002)

[9] Model-based Development: *"Model-based development refers to a development approach whose activities emphasise the use of models, tools and analysis techniques for the documentation, communication and analysis of decisions taken at each stage of the development lifecycle."* (El-khoury 2006)

Broad agreement exists between the authors in that the behavior of mechatronic systems cannot sufficiently be predicted before physical or virtual prototypes exist. Physical prototypes do not typically appear until later design phases where the relevant aspects of the mechatronic system were conceived in sufficient details. In contrast, the application of modeling and model analysis techniques enables predicting the system behavior already in early design phases (ID 2.6). Therefore the design methodology for mechatronic systems described in the VDI guideline 2206 (2004) comprises activities for modeling and model analysis reaching over the complete development cycle. Based on these models and simulation techniques, the fulfillment of system requirements concerning for instance static, dynamic, thermal, electric, magnetic, and other properties may be investigated at different stages of the design process. The term *"Model-based Development"* (MBD) subsumes all these activities centered on the broad application of models and model-based analysis techniques in product development (El-khoury 2006).

Table 1.4: Challenges within virtual prototyping and model-based development

Sources		Text-books	Engineering Guidelines		Research Reports		Dissertations		
		Krause et al. (2007), Innovationspotenziale	VDI 2206 (2004)	Angerbauer et al. (2010), aquimo	Aberdeen Group (2008), System Design	Eversheim et al. (2000), Chancen	Bellalouna (2009), Integrationsplattform	El-khoury (2006), Model Management	Kümmel (1999), Integration
ID	Challenges								
2.6	Prediction of the behavior of mechatronic systems not possible in early design phases without modeling and model analysis	●	●	●	●		●	●	●
2.7	Behavior of mechatronic systems can only be predicted based on a holistic model of the disciplinary models	●	●				●	●	
2.8	Phase-spanning use of models		●						
2.9	Reduction of physical prototypes and increase of virtual prototypes due to reduced development time and increased cost pressure				●				●

These days, model-based development is an integral part of the various disciplines involved in MPD. Even in software development, model-driven engineering technologies gain acceptance as they permit to (a) edit formal models using domain specific languages (DSL), (b) validate the correctness of the model based on quality criteria, and (c) generate source code, input for analysis tools, deployment descriptions etc. (Schmidt 2006). Unfortunately, the different disciplinary models describe the behavior of the mechatronic system strictly from their disciplinary points of view and contribute only partial aspects of the overall system behavior. For a complete analysis of the system behavior, it is therefore necessary to employ a multi-domain model describing the complete system behavior (ID 2.7).

In general, two categories of approaches for the design of such domain-spanning models may be distinguished:

(a) Multi-domain modeling approaches (Krause, Franke, et al. 2007) based on:
    i. Mathematical modeling spanning over multiple domains, as e.g. practiced in MATLAB/Simulink
    ii. Physical-modeling applying an object-oriented, multi-domain modeling language, e.g. Modelica (Andersson 1994)
    iii. Knowledge-based modeling: This modeling approach aims at explicitly including design knowledge and supporting the design process by context-sensitive mechanisms that supply the needed design knowledge (Bludau and Welp 2001)
    iv. Semantic modeling: Enrichment of models with semantic information in order to improve the capturing and provision of knowledge (Welp, Labenda, et al. 2007)
(b) Domain-spanning integration of disciplinary models based on:
    i. Mathematical models incorporating the domain-specific equations into a domain-spanning mathematical model (VDI 2004)
    ii. Meta-meta-models allowing for an integration of the tool specific data authored under their tool specific format (meta-model) (El-khoury 2006; ISYPROM Consortium 2011a)

The types of the applied models and design and analysis tools usually change over the different development phases. Consequently, the information embedded in models of previous development phases has to be reused in models of subsequent phases, i.e. a continuous flow of data and information has to be established in the chain of modeling and analysis tools (ID 2.8) (VDI 2004).

Two authors elaborate on the advantages for development cost and time arising from a trade of the number of physical prototypes for an increased number of virtual prototypes (ID 2.9). Virtual prototypes may be established faster and cheaper than physical prototypes (Boucher and Houlihan 2008). As a result, a higher number of product iterations may be developed and analyzed based on virtual prototypes, which additionally contributes to an improved product quality and performance.

1.1.2.3 Problems for Specification and Requirements Traceability

Table 1.5 gives an overview on problems arising from the needs for (a) holistic specification of mechatronic systems and (b) end-to-end tracking of requirements during MPD.

Two references state that the complexity of mechatronic products and the associated development process makes it difficult to keep track of the state of requirements throughout all iterations of the product (Kümmel 1999; Boucher and Houlihan 2008). Therefore, the problem of traceability[10] of design requirements needs to be addressed: Requirements have to be managed throughout the entire development process, which requires a consistent mapping of requirements to the various artifacts of the product structure existing in different development phases and disciplines, e.g. functions, logical elements, and physical subsystems (ID 2.10). Based on this consistent mapping of design requirements to items of the product structure, a superior understanding for the impact of design changes may be gained.

Table 1.5: Problems related to product specification and requirements traceability

Sources		Text-books		Engineering Guidelines		Research Reports		Dissertations	
		Krause et al. (2007), Innovationspotenziale	VDI 2206 (2004)	Angerbauer et al. (2010), aquimo	Aberdeen Group (2008), System Design	Eversheim et al. (2000), Chancen	Bellalouna (2009), Integrationsplattform	El-khoury (2006), Model Management	Kümmel (1999), Integration
ID	Challenges								
2.10	Traceability of design requirements over the entire product development process and to the different levels of the product structure				●				●
2.11	Domain-spanning specification of the mechatronic system	●	●						

The specification of the mechatronic system has to cover all involved disciplines (ID 2.11) (VDI 2004; Krause, Franke, et al. 2007). A simple and intuitive specification technique for the principle solutions of mechatronic systems is therefore needed, which allows for the domain-spanning specification of the system (Krause, Franke, et al. 2007). The inherent complexity of such a specification is illustrated by the fact that a complete specification of principle solutions for mechatronic systems includes no less than seven partial models spanning from requirements to environment (Frank 2006; cited in Krause, Franke, et al. 2007).

1.1.3 Structure Related Problems

After elaborating on the product and process related challenges in MPD, the following four sections focus on the various structural issues to be addressed in MPD. In particular, the identified literature sources will be analyzed with respect to challenges arising from (a) the structure of product data, (b) the structure of the IT landscape, (c) the structure of knowledge and

---

[10] Requirements Traceability: *"refers to the ability to describe and follow the life of a requirement, in both a forwards and backwards direction (i.e., from its origins, through its development and specification, to its subsequent deployment and use, and through periods of on-going refinement and iteration in any of these phases)."* (Gotel and Finkelstein 1994)

qualification, and (d) the structure of the development organization. Table 1.6 summarizes the structure related challenges.

1.1.3.1 Structure of Product Data

Each engineering disciplines involved in MPD typically applies its own product structure that is sufficient for its disciplinary purposes but not always compatible with the other disciplines. Additionally, these disciplinary product structures provide a taxonomy for the organization of the disciplinary product data which leads to redundancies and inconsistencies in the context of the overall mechatronic system (Bellalouna 2009). In design practice, however, there is a clear need for an up-to-date and real-time overview on the overall product structure and data during all development phases (Eversheim, Niemeyer, et al. 2000; Boucher and Houlihan 2008; Angerbauer, Buck, et al. 2010), which cannot be realized based on the heterogeneous product structuring and data. Therefore, MPD requires the definition and application of a multidisciplinary product structure (ID 3.1 in Table 1.6) (Boucher and Houlihan 2008), which may additionally provide the hierarchy for the organization of product data.

1.1.3.2 Structure of Information Systems

The IT landscape typically applied within MPD consists of various disciplinary tools and disciplinary data management systems that were in many cases originally developed to support specific activities in mono-disciplinary design. As a result, these tools were largely incompatible and could only be poorly integrated (VDI 2004). To overcome these limitations, integrated MPD environments (ID 3.2) have been proposed and conceived during the last decade (Eversheim, Niemeyer, et al. 2000; VDI 2004; El-khoury 2006; Bellalouna 2009) with the following characteristics:

(a) Integrated product data model spanning all disciplines and development phases providing the capabilities to map the tool specific data onto a common data model (El-khoury 2006; Bellalouna 2009)
(b) Coordination of the product data and information exchange (Angerbauer, Buck, et al. 2010), configuration and change management (Bellalouna 2009)
(c) Interdisciplinary collaboration capabilities (Boucher and Houlihan 2008)
(d) Multidisciplinary product data management and model management (El-khoury 2006)
(e) Initial description and management of requirements (VDI 2004)
(f) Systems engineering capabilities (VDI 2004; Boucher and Houlihan 2008)
(g) Functional modeling capabilities (VDI 2004)
(h) Integration of disciplinary modeling and simulations tools (VDI 2004)

1.1.3.3 Structure of Knowledge and Qualification

Mechatronic products emerge from the continuous interaction of broad knowledge originating from different disciplines. Each of the disciplines involved is rooted in its own terminology, knowledge base, processes, and modeling approaches (Krause, Franke, et al. 2007), which impedes the interdisciplinary collaboration required in MPD and in particular the cross-disciplinary flow of information and knowledge. In the context of knowledge management, Davenport and Prusak (2000) acknowledge this dilemma by stating that *"people can't share knowledge if they don't speak a common language"*.

In order to cope with the aforementioned difficulties of the interdisciplinary collaboration, MPD requires (among others) specific qualifications not needed in mono-disciplinary product development. On this subject, educational scientists have conducted research projects and gained insights; however the proper interpretation of these requires an understanding of the central concept of *"competence"*. In education science, the term *"competence"* has largely subsumed the concepts of knowledge, skill, qualification, and motivation over the last two decades. Weinert (1999) describes competence as *"a roughly specialized system of abilities, proficiencies, or individual dispositions to learn something successfully, to do something successfully, or to reach a specific goal."* More specific to the business sector, the term *"occupational competence"*[11] denotes the system of competences required for professional activities of individuals and collectives that displays itself in three distinct dimensions (Kultusministerkonferenz 2007; cited in Hackel 2010):

(a) Professional competence[12]: Willingness and ability to solve tasks and problems on the basis of professional knowledge and skills

(b) Social competence: Willingness and ability to create and shape social relations as well as to interact with others in a rational and responsible way

(c) Self-competence[13]: Willingness and ability to clarify and reflect on the personal opportunities, requirements, and constraints in the contexts of the family, professional and public life. It comprises personal traits such as autonomy, ability to take criticism, self-confidence, reliability, responsibility, and sense of duty.

Moreover, common competences such as (a) methodological competence, (b) communicative competence and (c) learning competence provide specific expertise to each of the aforementioned dimensions of occupational competence (Kultusministerkonferenz 2007).

Following the structure provided by the model of occupational competence, MPD necessitates - compared to mono-disciplinary product development - a range of new competences (ID 3.3) (Angerbauer, Buck, et al. 2010; Hackel 2010) :

(a) Professional competence:
  i.   Complexity handling: System thinking, structuring approaches (e.g. object-orientation, component orientation)
  ii. Problem solving: Knowledge on the mechatronic procedural model and on discipline-specific process models
(b) Methodological competence:
  i.   Mechatronic tools
  ii. Mechatronic methods applied within the mechatronic procedural model, e.g. for requirements analysis and system design
(c) Social competence:

---

[11] In German: „Handlungskompetenz"

[12] In German: „Fachkompetenz"

[13] In German: „Humankompetenz"

      i.   Interdisciplinary competence: Communication with experts from other disciplines, motivation for interdisciplinary cooperation, adaptation of different perspectives, interdisciplinary knowledge exchange

     ii.   Conflict handling competence

(d) Self-competence: self-organization, time management and learning competence

In addition to disciplinary competences typically provided by domain specialists, the handling of issues at the system-level, as e.g. for system design and system integration, requires generalist expertise usually contributed by system engineers (Kühnl 2007). In general terms: The knowledge possessed by a single person will not be sufficient to cover the full width and depth of knowledge required in MPD. Therefore, (a) multiple team members with knowledge from different domains and (b) several organizational units typically contribute to the overall knowledge applied during the development process. This results in widespread and complex networks of information and knowledge exchange spanning across disciplinary and organizational boundaries (Neumann 2012) (ID 3.4).

The success of product development does not only depend on the initial knowledge existing once product development is started, furthermore it is linked to the knowledge acquired during development through activities as e.g. experimentation (Avgoustinov 2007) in MPD typically based on modeling and model analysis methods. From a knowledge perspective, the model-based development approach practiced in MPD therefore offers significant advantages as it allows to (a) decrease the costs of knowledge acquisition and (b) increase the learning speed and amount of acquired knowledge (Avgoustinov 2007).

These characteristics of the model-based approach as well as the numerous product iterations typically conducted in MPD result in high volumes of acquired knowledge (Neumann 2012) that have to be properly managed and dispatched. Consequently, of these acquired pieces of information and knowledge, the aspects relevant to each of the different disciplinary and organizational participants need to be properly identified, then transformed in a form comprehensible to the particular recipient, and lastly distributed to these stakeholders (Neumann 2012). The high knowledge dynamics of MPD, compared to sequential and mono-disciplinary product development, necessitates an enhanced support for the creation, storage, distribution, and application of knowledge (Neumann 2012) (ID 3.5) comprising the operative side of knowledge management (Bodendorf 2006).

The decisive role of knowledge management activities is highlighted in a study on *"Innovation strategies of successful automotive suppliers"* (Roth 2008) by the fact that 90% of successful automotive suppliers implement or already practice knowledge management whereas this applies only to 41% of the less successful suppliers.

1.1.3.4 Structure of Organization

Historically, development organizations have been structured into e.g. mechanical and electrical/electronic departments according to disciplinary aspects. The transition to interdisciplinary product development in Mechatronics induces increased cooperation and coordination requirements between the disciplinary departments, which are not well supported by the existing organizational structure (Eversheim, Niemeyer, et al. 2000). Consequently, an organizational integration and transformation of mechanical, electronic, and software development departments toward a product-oriented Mechatronics department provides a better suiting organizational structure for MPD (ID 3.6) (Eversheim, Niemeyer, et al. 2000; Angerbauer, Buck, et al. 2010).

In the context of project management for MPD, interdisciplinary teams are widely considered as the optimal project organization (Kümmel 1999; Eversheim, Niemeyer, et al. 2000; VDI 2004) due to better knowledge coverage and resulting higher creativity (ID 3.7).

Table 1.6:　Structure related challenges

Sources		Arti-cles	Engineering Guidelines		Research Reports		Dissertations		
		Neumann (2012) Potentials, Challenges	VDI 2206 (2004)	Angerbauer et al. (2010), aquimo	Aberdeen Group (2008), System Design	Eversheim et al. (2000), Chancen	Bellalouna (2009), Integrationsplattform	El-khoury (2006), Model Management	Kümmel (1999), Integration
ID	Challenges								
3.1	Definition and application of multidisciplinary product structure				●				
3.2	Integration of various disciplinary tools within integrated MPD environments		●			●	●	●	
3.3	MPD necessitates a range of new competences			●					
3.4	Widespread and complex networks of information and knowledge exchange spanning across disciplinary and organizational boundaries	●							
3.5	Model-based approach and numerous product iterations lead to high knowledge dynamics	●							
3.6	Disciplinary organization structures impede the increased cooperation and coordination requirements in MPD			●		●			
3.7	MPD requires interdisciplinary teams		●			●			●

1.2 Problem Definition, Objectives, and Research Questions

In general, scientific research requires a clearly structured approach to produce research results in an efficient and truly scientific manner. For this purpose, a research methodology defines the research approach as well as the supporting methods and guidelines within a particular scientific discipline. Within the area of design research, however, Blessing and Chakrabarti (2009) stated a lack of consensus on research methods and on a common research methodology. Reich (1995) describes the commonly followed pattern in design research that research apprentices simply learn the research method from their senior peers without discussing sufficiently its underlying assumptions and associated limitations.

For these reasons, Blessing and Chakrabarti (2009) established a research methodology called *"Design Research Methodology"* (DRM) drawing from (a) the categorization of research literature provided by Finger and Dixon (1989a; 1989b), (b) the research approaches practiced in other disciplines, and (c) the previous publications describing approaches to design research. Since the middle of the 1990s, the precursors of DRM have been followed by Blessing and Chakrabarti's students. In addition, the methodology has been taught in the Summer School on Engineering Design Research beginning in 1999. Due to this long and positive record of accomplishment of DRM, the present thesis also adopts the DRM approach.

DRM recommends the formulation of a research plan essentially comprised of the research problems to be addressed, the research goal, and main research questions. Following this research methodology, the present thesis lies out the main constituents of the research plan in the next two sections.

The literature review in the previous section uncovered numerous challenges for MPD, which have been the subject of scientific research since 1999. Meanwhile, research has already addressed several of the presented challenges in MPD up to the point where e.g. dedicated methodological guidelines, modeling approaches, or development tools materialized and found increasing application in the industry as shown by the following selection of accomplishments:

(a) Modelica (an object-oriented multi-domain modeling language) addresses several of the raised issues in the area of multi-domain modeling and virtual prototyping of complex systems (IDs 2.3, 2.6, 2.7, 2.9). So far, support for Modelica has been implemented in nine modeling tools and it is entering mainstream industry usage in system design (Casella 2009).

(b) Recent releases of enterprise PDM systems (e.g. ENOVIA v6, Teamcenter, Windchill) added support for domain-spanning processes for e.g. integration, configuration, and change management (ID 2.5).

(c) CATIA V6 Systems, one of the recently appeared integrated Mechatronics development environments, addresses problems of the phase-spanning use of models (ID 2.8).

(d) Two of the integrated Mechatronics development environments (CATIA V6 Systems, Siemens PLM Mechatronics Concept Designer) target issues in the area of specification and requirements traceability (IDs 2.10 and 2.11).

(e) Recent versions of enterprise PDM systems (e.g. ENOVIA v6, Teamcenter, Windchill) aim at the handling of multidisciplinary product structures (ID 3.1).

(f) At least three integrated Mechatronics environments (ID 3.2) appeared in recent years and start to enter industry usage.

(g) Numerous universities developed programs for Mechatronics engineering (Wikander, Törngren, et al. 2001; Grimheden and Hanson 2005) that are aimed at educating engineers with an well-suited competence profile for MPD (ID 3.3).

In contrast, some others of the described challenges in MPD as e.g. product complexity and technological heterogeneity (IDs 1.1, 1.2) remain in the focus of scientific research. Among these currently research-relevant topics, the present thesis selects such issues that on the one hand bear a significant potential for improvement within MPD, and that, on the other hand, have not been in the focus of researchers and the industry in the past. Therefore, it selects the following two topics as its main research problems (Neumann 2012):

(a) Coordination and collaboration hindered by differences in terminology, methods and procedures between disciplines (ID 2.2)

(b) Widespread and complex networks of information and knowledge exchange (ID 3.4) and high knowledge dynamics (ID 3.5) call for enhanced support for the creation, storage, distribution, and application of knowledge

The following research objectives will address the aforementioned research problems in a stepwise approach (Neumann 2012):

1. Clarification of the terminology associated with MPD and the nature of cross-disciplinary interactions in MPD and Mechatronics engineering science

2. Clarification of the terminology for data, information, and knowledge, supporting knowledge technologies and approaches in the context of product development

3. Conception of a research framework supporting the analysis of knowledge bases, knowledge structures, and knowledge flows in product development

4. Identification of common process elements of MPD

5. Analysis of the knowledge characteristics of these process elements based on the aforementioned research framework

6. Research on semantic technologies allowing for a semantic enrichment of the various models used in MPD in order to improve the processing, storage, distribution, and context-sensitive provisioning of knowledge in MPD

7. Verification of the conceived design support system for MPD based on an application scenario

Blessing and Chakrabarti (2009) emphasize the twofold objective of design research that not only aims to increase the understanding of design, but also focuses on improving design. The second objective is typically realized by a particular design support aiming at developing the existing situation into the desired situation (Blessing and Chakrabarti 2009). In order to improve the handling of knowledge in MPD, the present thesis aims to develop a design support system with the following characteristics:

(a) Semantic enrichment of models used in MPD in order to extract and store their embedded knowledge

(b) Context-sensitive provisioning of knowledge for the different disciplines and roles in MPD

According to the DRM methodology, research questions are a helpful means to express the research issues that need to be addressed. In this spirit, the present thesis aims to provide the answers to the following research questions:

1. What is the most appropriate disciplinary interaction model describing the nature of the cross-disciplinary collaboration in MPD and in Mechatronics engineering science?
2. How might knowledge constituents relevant within product development (knowledge resources, structure of knowledge and knowledge flows) be analyzed? What could be the appearance of a research framework that provides the methodological basis for the analysis of the knowledge characteristics of product development?
3. Which process elements are commonly applied within MPD?
4. What are the knowledge needs of the different disciplines and roles within each of these process elements?
5. What are adequate semantic technologies permitting a semantic enrichment of the numerous types of models used within MPD allowing to improve the processing, storage, distribution and context-sensitive provisioning of knowledge?
6. What is an appropriate system architecture for a design support system based on semantic technologies aiming at improving the processing, storage, distribution, and context-sensitive provisioning of knowledge within MPD?

1.3 Approach and Structure of the Thesis

Figure 1.4 depicts the approach and structure of the thesis and relates its chapters to the stages as defined by DRM (Blessing and Chakrabarti 2009), the overall applied research methodology of this thesis.

The first chapter gives an introduction on the current situation and challenges of MPD, which provides the motivation for the research conducted in the present thesis. Next, a review of the relevant literature on MPD since 1998 provides the input for the collection, structuring, and systematization of problems encountered in MPD. Based on the uncovered research-relevant topics in MPD, the present thesis selects a few of these issues as its main research problems. Then, the remaining parts of the research plan, i.e. research objectives and main research questions, are laid out. According to DRM, the first chapter constitutes the *"Research Clarification"* stage.

The second chapter aims at the clarification of the relevant terminologies associated with MPD, Mechatronics engineering science, and mechatronic systems. In addition, it clarifies the nature of cross-disciplinary interactions in MPD and Mechatronics engineering science.

The third chapter focuses on the clarification of the concepts of data, information and knowledge as well as related processes. Next, it introduces selected taxonomies for the structuring of knowledge.

Chapter 4 gives an overview on cognitive psychology and on various subfields of thinking and reasoning. Drawing from well-recognized dual process theories for the various subfields of thinking, it introduces the tripartite model of mind that will be applied throughout the present thesis.

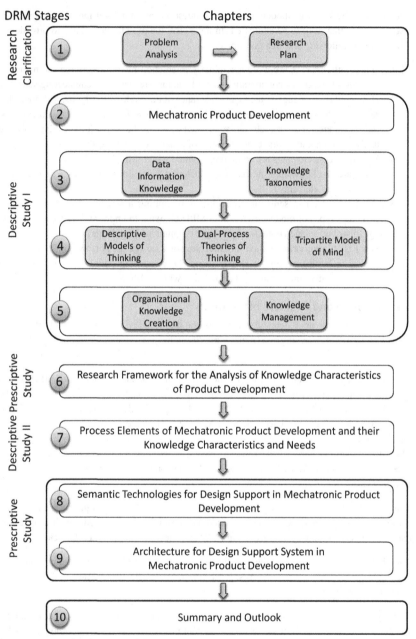

Figure 1.4: Approach and structure of thesis

The fifth chapter provides an introduction on the model for organizational knowledge. Next, it introduces the aims, components, and strategies of knowledge management.

Chapters 2 to 5 form a *"Descriptive Study I"* in terms of DRM that aims at gaining a deeper understanding of the current situation in the relevant research fields.

The sixth chapter develops a descriptive model of knowledge creation in interdisciplinary product development. Next, it introduces specific knowledge taxonomies specialized to product development and describes the conception of a research framework aiming at the analysis of knowledge resources, knowledge contexts, and knowledge flows in product development. It constitutes a *"Prescriptive Study"* in the context of DRM that aims at conceiving a research framework for the analysis of knowledge characteristics of product development.

Chapter 7 applies the aforementioned research framework to MPD. In particular, it identifies process elements common to Mechatronic product development and analyzes the knowledge characteristics and knowledge needs of these process elements. In terms of DRM, it forms a *"Descriptive Study II"*.

The eighth chapter collects the high-level requirements for a design support system aiming at improving the processing, storage, distribution, and context-sensitive provisioning of knowledge in MPD. Next, it looks for appropriate semantic technologies allowing for a semantic enrichment of the various models used in MPD.

Chapter 9 develops the system architecture for the intended design support system for MPD based on the semantic enrichment of models and the context-sensitive provisioning of information and knowledge for the different disciplines and roles in MPD.

Chapters 8 to 9 form a *"Prescriptive Study"* in the context of DRM that aim at establishing the intended design support based on a dedicated research framework and the previously identified common process elements of MPD.

Finally, chapter 10 summarizes the findings of the thesis, answers the research questions laid out in the first chapter, and elaborates on future research needs uncovered by the thesis.

2 Mechatronic Product Development

The world of engineering is like an archipelago whose inhabitants are familiar with their own islands but have only a distant view of the others and little communication with them. A comparable near-isolation impedes the productivity of engineers, whether their field is electrical and electronics, mechanical, chemical, civil, or industrial. Yet modern manufacturing systems, as well as the planes, cars, and computers, and myriad other complex products of their making, depend on the harmonious blending of many different technologies.

Richard Comerford, 1994

In the quote above, Comerford (1994) tellingly depicts the situation which initiated the development of Mechatronics. Characterized by the cited blend of technologies, Mechatronic products have gradually emerged from electromechanical products through the addition of first electronics, later computer controls, and software. Ko Kikuchi, the former president of the Yasakawa Electric Company, introduced the term *Mechatronics* from *"mecha"* for mechanism and *"tronics"* for electronics in 1969, in order to use the term as trademark for such devices (Comerford 1994).

Precision mechanics was the initial field where Mechatronics emerged (Diehl 2009). These days, Mechatronics encompasses a much larger scope of applications. It ranges from mechatronic components (e.g. semi-active hydraulic shock absorber, integrated servo drive) over mechatronic machines (e.g. mechatronic combustion engines, integrated AC drive systems) to the omnipresent applications in the automotive area (e.g. anti-lock braking systems, electrohydraulic brakes, active suspension systems) (Isermann 2008) and in the consumer goods industry (e.g. vacuum cleaners, washers, dryers, dishwashers, entertainment devices) (de Silva 2005).

2.1 Introduction and Terminological Understanding

Over the years, though, the usage of the term Mechatronics became increasingly heterogeneous (Zohm 2003). A rough analysis of the about 20 definitions on Mechatronics collected at the website of the Department of Mechanical Engineering at Colorado State University (Alciatore 2011), uncovers large differences on the contextual level of the definitions:

(a) Two thirds of the definitions refer to the context of design methodology and locate Mechatronics in between a design approach, design methodology, and design framework.
(b) The remaining third of the definitions views Mechatronics primarily as an engineering discipline.

(c) Only one author perceives Mechatronics from the context of the conceived technical systems.

This controversy about substantial aspects of Mechatronics is reflected in the opinion of several authors who state that a generally accepted definition of Mechatronics does not currently exist (Buur 1990; Hewit 1993; cited in Wikander, Törngren, et al. 2001; VDI 2004; Bishop 2006; Felgen 2007; Vajna, Weber, et al. 2009). Moreover, Hewit (1993; cited in Wikander, Törngren, et al. 2001) argues that such a definition should not be desirable, as it would constrain the future development of this engineering field. Correspondingly, Bishop (2006) states that the lack of consensus on a widely accepted definition should be interpreted as a healthy sign for a vivid engineering field. He endorses a broad perception of Mechatronics as *"a natural stage in the evolutionary process of modern engineering design"* (Bishop 2006).

2.1.1 Contexts of Terminological Understanding

In spite of the ongoing debate on the nature of Mechatronics, the present thesis is in need of a clear terminological understanding of Mechatronics in order to conduct research on knowledge characteristics at the core of MPD, i.e. its knowledge bases, the used knowledge frame and the knowledge dynamics tightly associated with the manifold activities during the product development process.

The aforementioned analysis of the contextual levels used for definitions of Mechatronics (Alciatore 2011) indicates that Mechatronics is perceived within the scientific world from different points of views, including at least three distinct contexts:

    (a) Mechatronics engineering science
    (b) The approach and methodology practiced in MPD
    (c) Mechatronic systems

When using the similar term *electronics* as analogy, it appears that this term likewise operates on at least two different contextual levels: Firstly, it designates a *"branch of physics and technology concerned with the design of circuits using transistors and microchips, and with the behavior and movement of electrons in a semiconductor, conductor, vacuum, or gas"* (Oxford Dictionaries 2010). Secondly, it describes the products of these design activities, i.e. *"circuits and devices using transistors, microchips"* (Oxford Dictionaries 2010).

In compliance with the results of the analysis of numerous definitions on Mechatronics and the aforementioned analogy to electronics, the present thesis regards Mechatronics as a concept appearing within a range of contextual levels (cf. Figure 2.1):

    (a) The practice of mechatronic product development will be considered as the primary context for the definition of Mechatronics.
    (b) In addition, Mechatronics engineering science will be taken into account as a second context because it acquires and supplies a large part of the knowledge applied within MPD and contributes therefore substantially to MPD's knowledge characteristics.
    (c) Equally, a terminological understanding of Mechatronics in the context of mechatronic systems will be beneficial due to the tight relationships between (i) the product's

specific content and structure, and (ii) the characteristics of the product development process itself.

Figure 2.1: Contexts for the terminological understanding of Mechatronics (Neumann 2012)

In the following, these three contexts of terminological understanding will be explored to (a) extract key aspects shaping the nature of Mechatronics applicable to all of these contexts, (b) discover topics in each context closely connected to these paradigms, and finally (c) provide definitions of Mechatronics within each of the contexts comprising those specific topics.

2.1.2 Paradigms of Mechatronics

Zohm (2003) analyzed in his dissertation the scientific understanding of Mechatronics based on definitions given in text books, dissertations and journals, and arrived at the following insights:

(a) There is a large agreement between the authors on the interdisciplinary character of Mechatronics, which Zohm interprets plainly as *"collaboration of several disciplines"*.

(b) The majority of publications mention three disciplines as constituents of Mechatronics: mechanics, electronics, and computer science. Other authors subsume additional disciplines, e.g. control engineering, or perceive nearly all engineering disciplines as constituents of Mechatronics.

(c) Only a few publications provide a sufficient differentiation of the integration into functional and spatial aspects. Zohm emphasizes the distinction of functional and spatial integration as central element for the terminological understanding of Mechatronics.

In contrast to the findings of Zohm (2003), the analysis of the about 20 definitions listed on the aforementioned website (Alciatore 2011) confirms that control engineering is considered by most definitions as an essential contributor to Mechatronics. Therefore, the present thesis takes into account mechanical engineering, electrical engineering, and computer science with control engineering as one of its sub-disciplines, as the main constituents of Mechatronics as shown in Figure 2.2 (VDI 2004; Isermann 2008; Diehl 2009).

Other scholars consider Mechatronics rather as a *"design philosophy"* (Millbank 1993; Tomkinson and Horne 1996; cited in Grimheden and Hanson 2001). Certainly, that designation of Mechatronics as *"design philosophy"* collides with the established meaning of design philosophy within engineering design science where it is considered *"as a meta-theoretical framework for design theories"* (Love 2000; cited in Horváth 2004). In order to avoid this contradiction, the word *"design philosophy"* should be interpreted based on its second meaning[14] as a particular approach to engineering design combined with a specific mechatronic attitude.

Machine Design
Precision Mechanics
Electro-Mechanics
Fluid Mechanics

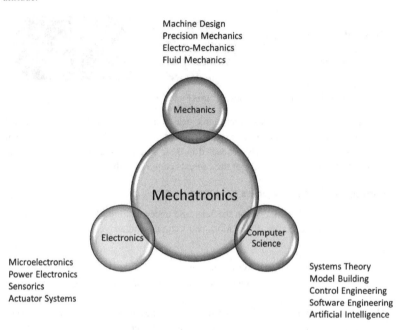

Microelectronics
Power Electronics
Sensorics
Actuator Systems

Systems Theory
Model Building
Control Engineering
Software Engineering
Artificial Intelligence

Figure 2.2: Main constituents of Mechatronics, adapted from (Isermann 2008)

What constitutes the essence of that specific mechatronic attitude or approach? Grimheden (2005) emphasizes in his dissertation on *"Mechatronics engineering education"* the central importance of the *"concept of synergy, or synergistic integration"* for the thematic identity of Mechatronics.

Harashima et al. (1996) introduced a closely associated definition of Mechatronics, which serves as the central reference within the VDI guideline 2206:

"[Mechatronics is]... the synergetic integration of mechanical engineering with electronic and intelligent computer control in the design and manufacturing of industrial products and processes."

Bradley (2010) discusses in his article *"Mechatronics – More questions than answers"* the numerous open questions on the nature of Mechatronics after 40 years of its evolution and the future potential of Mechatronics. He questions the uniqueness of the MPD approach given the fact that now the majority of engineering systems are integrated and interdisciplinary and concludes *"that mechatronics has to a significant degree been integrated with and incorpo-*

---

[14] The Oxford Dictionaries (2010) support this interpretation by providing a second meaning of philosophy as *"a theory or attitude that acts as a guiding principle for behavior"*.

rated within mainstream engineering design methods and strategies" (Bradley 2010). In addition, he raises the question whether Mechatronics today should rather be considered as a systems oriented design approach.

The key aspects of Mechatronics described in the definitions given by Harashima et al., the dissertations of Buur, Zohm and Grimheden, i.e. synergy, spatial and functional technology integration, interdisciplinary character, and the understanding of Mechatronics as a systems oriented approach to engineering design (as shown in Figure 2.3), provide the input for the overall terminological understanding of Mechatronics within the present thesis.

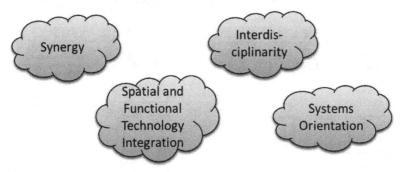

Figure 2.3: Key aspects shaping the nature of Mechatronics

2.1.3 Definitions of Mechatronics

Based on the described paradigms of Mechatronics, the sections 2.3, 2.4 and 2.5 will elaborate on specific fields of activities in the contexts of mechatronic systems, MPD and Mechatronics engineering science, which will then serve as the foundation for a definition of Mechatronics in each of these areas.

2.2 The Evolution of Mechatronic Products

Eversheim et al. (2000) introduced an evolutionary model describing the gradual changes from purely mechanical, through electro-mechanical, to mechatronic products in four phases. This model was later refined and extended by Zohm (2003) and finally consists of five development stages. It depicts in detail the basic activities of technological change, e.g. substitution, functional and spatial integration, leading to upgraded technical solutions for the initially purely mechanical product. The model allows for a flexible description of change, as it permits the omission of specific development phases not applicable for particular types of products. In the following, the evolutionary model is presented based on the descriptions given in Zohm's dissertation (Zohm 2003).

Figure 2.4 depicts the evolution of an automotive window-lifting system as an example illustrating the activities and phases of the evolutionary model. The initial product consists of purely mechanical components that realize all product functions (*Phase I*).

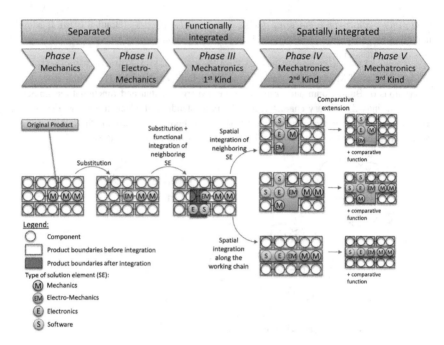

Figure 2.4:   Evolutionary model of mechatronic products applied to the evolution of an automotive window lift
system, according to (Zohm 2003)

In the next phase (*Phase II*), electro-mechanical components, e.g. relays, electric motors and
other electric actuators, complement or replace the existing mechanical elements. In the case
of the automotive window-lifting system, an electric motor controlled by a switch replaced
the previous manual crank handle in the early 1940s in a few luxury car models (Society of
Automotive Engineers 1940).

The addition of electronic components, e.g. microprocessors, transistors, thyristors and sensors, allows for new and upgraded product features in the subsequent evolutionary phase
(*Phase III*). Typically, a separate printed circuit board contains these electronic components.
Wired connections link the board with functionally related components. Due to demanding
constraints on durability toward mechanical loads and thermo-dynamic conditions, electronics
cannot yet be integrated spatially with the neighboring electro-mechanical and mechanical
components.

Programmable electronic components, typically a microprocessor, permit the execution of a
dedicated software implementing functions such as information processing, decision making
and control of the actuators. As a result, embedded software becomes part of the mechatronic
product. The development of this first kind of mechatronic systems requires therefore the contributions of all main engineering disciplines constituting Mechatronics (cf. section 2.1.2).

As this stage primarily aims at achieving a functional integration of the solution elements, an optimization of the system reaching over all involved disciplines is not yet in scope. The collaboration of the involved engineering disciplines may therefore remain on the multidisciplinary level. In the 1980s, such so called *"first kind of mechatronic systems"* allowed for new features within automotive window-lifting systems such as pinch protection (Brose Fahrzeugteile 2006).

Phase IV focuses on the spatial integration of the already functionally integrated components, which requires a closer, interdisciplinary collaboration addressing the challenging mechanical, thermo-dynamic and spatial conditions for the integration of the heterogeneous components. In the example of the window-lifting systems, the spatial integration of the electric motor, the gearbox, position sensors, and the connector for the associated control unit within one casing increased the density of functions per volumetric unit, simplified the assembly process, and allowed for a flexible usage of different types of control units fitting into an appropriate slot. As a next step in the integration process, so called *"door systems"* integrated most of the components equipping a vehicle door, e.g. the window lift, the door-locking system, loudspeakers, the wiring harness, and the control unit. As of 2006, door systems already comprised 85% of the functional components of a vehicle door (Brose Fahrzeugteile 2006).

The fifth development stage of mechatronic products (*Phase V*) focuses on the further optimization of the already functionally and spatially integrated solution elements toward the generation of additional synergistic effects. At this stage, these spatially integrated products may permit the realization of upgraded and new functions not achievable in earlier development phases of the evolution process. In case of the window-lifting system, the fifth stage resulted for instance in a more reliable and cost-effective pinch protection system based on the spatial integration of two hall sensors into the electric motor (Reif 2009). Zohm (2003) argues that these new functions of spatially integrated mechatronic systems may permit companies to achieve a comparative advantage[15] over their competition. Therefore, he introduces the term *"comparative functions"* to indicate functions enabling such comparative advantage.

2.3 Mechatronic Systems

2.3.1 Characteristics and Definition

The preceding section clarified how technological activities shaped the characteristics of mechatronic products during an evolutionary process. In the following, the specific impact of these technological activities on the content, the structure, and the capabilities of mechatronic systems will be described and utilized to define the essence of mechatronic systems.

---

[15] Comparative advantage: The ability of an organization to perform an economic activity more efficiently than others (Oxford Dictionaries 2010).

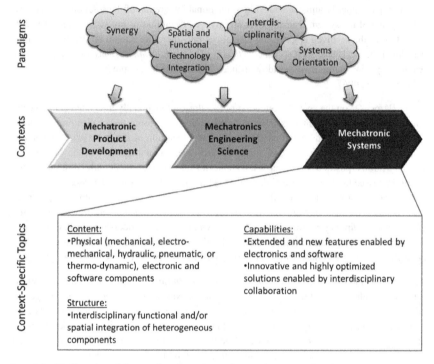

Figure 2.5:   Manifestations of the overall mechatronic paradigms in the context of mechatronic systems

Figure 2.5 depicts the resulting characteristics of mechatronic systems:

(a) Their content became technologically heterogeneous.

(b) Due to functional and/or spatial integration, their structure is characterized by the high number of relations between items belonging to different disciplines.

(c) Their capabilities were extended under the influence of electronics and software.

(d) All together, they have the potential to constitute innovative and highly optimized solutions.

These characteristics of mechatronic systems, the understanding of a mechatronic system as a specialized technical system according to the definition given by VDI guideline 2221 (1993), and the definition of mechatronic systems provided by Isermann (2008) present the essential inputs for the conceptual understanding of mechatronic systems. The following definition integrates these different contributions into the terminological concept adopted by the present thesis:

Definition 2.1:   Mechatronic system

A mechatronic system is a technical system that consists of functionally and/or spatially integrated physical (mechanical, electro-mechanical, hydraulic, pneumatic, or thermodynamic), electronic and software components. The integration of these heterogeneous components is achieved first by a multidisciplinary and later by an interdisciplinary collaboration of the relevant engineering disciplines (typically: mechanical engineering, electrical engineering, and computer science with control engineering as one of its subdisciplines) that enable synergistic effects resulting in innovative and highly optimized solutions.

2.3.2 Structure of Mechatronic Systems

2.3.2.1 Basic Structure

In principle, four main components integrated with each other will form the basic structure of a mechatronic system (VDI 2004; Gausemeier and Feldmann 2006).

The *basic system* constitutes in the most general sense a physical system, whose state is continuously observed and controlled. Using the example of an automotive window-lifting system (cf. section 2.2) using cables, a so called *"cable window-lifter"*, the basic system contains only mechanical components. Here, a gearbox, the cable, and the lifting arms moving the window glass form the basic system. In more general terms, the basic system may comprise *"a mechanical, electro-mechanical, hydraulic or pneumatic structure or a combination of these"* (VDI 2004).

Sensors measure the state variables of the basic system and transform the measured values into electric quantities serving as input for the information-processing unit. With the automotive window-lifting system, a simple pinch protection system may be realized solely based on software sensors – the monitoring of the alternating component of the motor current permits the calculation of the current window position. A more reliable pinch protection approach requires physically present sensors: Two hall sensors measure the motor's rotation speed and direction permitting the calculation of the current displacement force (Reif 2009).

The *information processing* handles the data supplied by the sensors and decides on this basis how the actuators have to be controlled in order to influence the basic system in the preferred direction. These days, a microprocessor which only handles digital data constitutes this information-processing unit. For the automotive window-lifting system, the microcontroller is spatially integrated with other electronic components (e.g. EEPROM, voltage regulator, analog-digital converter, LIN-transceiver) on a single application specific integrated circuit (ASIC) (Reif 2009).

Controlled by electric actuating variables issued by the information-processing unit, the *actuator* influences the basic system typically by mechanical energy toward the desired state. In the example of the automotive window-lifting system, an electric motor constitutes the actuator.

Figure 2.6 depicts the three types of flows (Roth 1994; VDI 2004) linking the mechatronic system's main components:

(a) *Information flows* (I) carry different forms of information between the system's components, e.g. values measured by sensors, signals issued by switches and control information driving the actuators.

(b) *Energy flows* (E) transport the different types of energy (e.g. mechanical, electrical, thermal, and magnetic) between the units of the mechatronic system.

(c) *Material flows* (M) realize the exchange of material between the components of the mechatronic system, e.g. solid bodies, liquids, and gases.

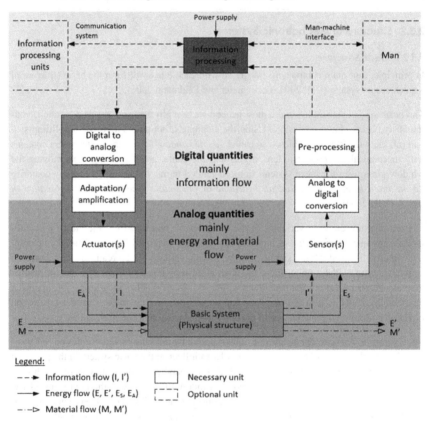

Figure 2.6: Basic structure of a mechatronic system, adapted from (Gausemeier, Ebbesmeyer, et al. 2001; cited in Krause, Franke, et al. 2007)

In the upper area of Figure 2.6, the information-processing unit, sensors and actors exchange mainly information typically based on digital quantities. In addition, optional components such as a man-machine interface and external information-processing units may be connected with the information processing. Energy and material flows dominate the relationships of the components in the lower part of Figure 2.6.

2.3.2.2 Modularization and Hierarchization

Based on the understanding of the aforementioned basic structure of a mechatronic system as building block for complex mechatronic systems, Lückel (2000) introduced a hierarchical structuring of mechatronic systems into three distinct layers (cf. Figure 2.7) where the basic structure referred to as mechatronic functional module (MFM) builds the first level.

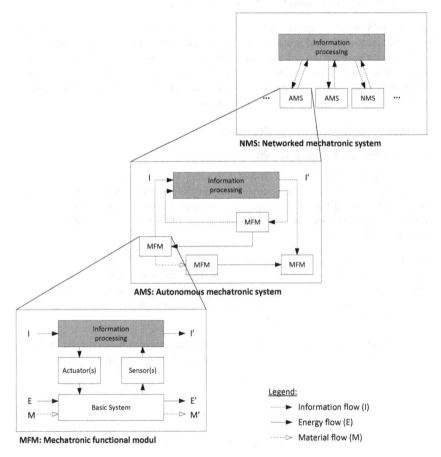

Figure 2.7: Structuring of mechatronic systems, according to (Lückel, Koch, et al. 2000; cited in Krause, Franke, et al. 2007)

At the second level, multiple MFMs linked by information flows and/or flows of mechanical energy form an autonomous mechatronic system (AMS) (Krause, Franke, et al. 2007). Besides the coupled MFMs, the AMS comprises its own information-processing unit and optional sensors. Functionally, an AMS focuses at the realization of high-level tasks, e.g. error diagnosis and error monitoring, as well as the delivery of set points for subordinate MFMs (VDI 2004).

An autonomous transport robot, used as transport system in e.g. warehouses and factories, comprises several MFMs, such as the drive system consisting of the wheel drives, sensors capturing the wheel's rotation speed, sensors dedicated to collision detection, and a microcontroller. At the level of the AMS, the information-processing unit focuses on high-level tasks, as e.g. the communication with the coordinator unit of the superior hierarchy level on current and future transport jobs, and the subsequent planning of these transport jobs.

Multiple AMS linked through information flows build up the third hierarchy level – they form a so-called networked mechatronic system (NMS). Here, the information processing of the NMS realizes coordination tasks for the AMS and NMS comprising this level (Krause, Franke, et al. 2007). In case of the robot-based transport system, the information processing of the NMS handles e.g. incoming transport jobs, dispatches them to the available transport robots, and keeps track of the fulfillment of these tasks.

2.4 Mechatronic Product Development

2.4.1 Characteristics and Definition

In the context of purely mechanical products, the term *"design"* or more precisely *"engineering design"* covers all engineering activities targeting the development of technical systems and products, starting from a *task definition*, going over the *elaboration of* the *information* required for the production and usage of the product, and finally leading to the *product documentation* (VDI 1993).

With the change toward mechatronic products and a broader, multidisciplinary perspective in the development of these products, the term *"design"* was replaced by *"product development"* (Vajna, Weber, et al. 2009), comprising both the interdisciplinary and the different disciplinary development process steps. In this spirit, the term *"product development"* is applied throughout the thesis in the context of MPD.

Figure 2.8 compiles and depicts the insights gained in previous sections on specific challenges, approaches, and overall characteristics of mechatronic product development:

(a) The development approach in MPD is characterized as integrated, interdisciplinary, and systems oriented.
(b) On the technological side, synergies arise from the integration of mechanics, electronics, control engineering, and software.
(c) MPD extensively uses the principles and tools of model-based development.

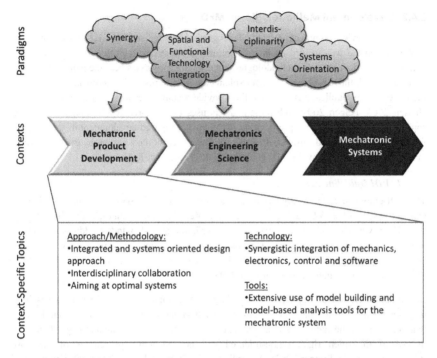

Figure 2.8: Manifestations of the overall mechatronic paradigms in the context of mechatronic product development (Neumann 2012)

The terminological understanding of MPD consists of the aforementioned characteristics of MPD as well as the more apparent understanding of MPD as product development specialized to mechatronic systems. The following definition outlines this terminological concept:

Definition 2.2: Mechatronic product development (Neumann 2012)

Mechatronic product development comprises the integrated engineering activities targeting the development of mechatronic systems. It involves an interdisciplinary collaboration typically between mechanical engineering, electrical engineering, and computer science with control engineering as one of its sub-disciplines. It applies a systems oriented design approach, i.e. it considers the objects of design activities as technical systems and adopts procedures, methods, and tools of systems engineering accordingly. Additionally, it follows the principles of model-based development, i.e. it broadly uses models, model-based tools, and analysis techniques to support its different activities.

2.4.2 Development Methodologies in MPD

The literature review on process-related challenges in MPD (cf. section 1.1.2) listed numerous challenges on the way toward an interdisciplinary product development process. Here, domain-spanning development methodologies and reference processes bear the potential to cope adequately with those problems, which explains the intense research interest in this field. As there are already publications available that provide a comprehensive review on this subject (Jansen 2006; Felgen 2007), the present thesis uses the research results laid out in these sources. Therefore, a detailed presentation of the many proposed development methodologies for MPD will be omitted. Instead, it focuses on the currently most comprehensive procedure model for MPD laid-out by the VDI guideline 2206 (Felgen 2007).

2.4.2.1 VDI Guideline 2206

The VDI guideline 2206 *"Design Methodology for Mechatronic Systems"* aims at filling the gaps left by the older VDI guidelines 2221 and 2422. These guidelines, dating back to the early 1990s when MPD had a comparatively lower importance than today, address the specific needs of interdisciplinary product development only partially. Likewise, they could not take into account the results of the intensified Mechatronics-oriented research in the 1990s and thereafter (Gausemeier and Möhringer 2003).

In addition to the overall objective of providing methodological support for the interdisciplinary development of mechatronic systems (VDI 2004), the guideline draws from the finding that engineers in the industry require flexible and situation-specific methodological support instead of the often rigid, phase-oriented and not adaptable procedures of the past (Gausemeier and Möhringer 2003). Accordingly, the VDI guideline 2206 (2004) proposes a flexible procedure model that consists of three main constituents and two extension points where the latter ones supply the required flexibility toward specific situational conditions:

    (a) Main constituents for design support:
        i.   The cycle of problem solving at the micro-level
        ii. The V-shaped procedure model on the macro-level
        iii. Pre-defined process modules for recurring design activities
    (b) Extension points for the customization of procedures:
        i.   Adaptation of pre-defined process modules or creation of new user-specific ones
        ii. Variation of the number of cycles on the macro-level (product iterations)

In the following, the cycle of problem solving, the procedure model and associated process modules are presented based on the descriptions given in VDI guideline 2206 (2004).

2.4.2.1.1 Micro-Level: Cycle of Problem Solving

The cycle of problem solving (cf. Figure 2.9) constitutes a general approach for the handling of tasks and the solving of problems as they have typically to be addressed within product development. The approach originates in systems engineering and has proven to be efficient in many different disciplines (VDI 2004).

The cycle may be either initiated with (a) a *situation analysis* based on the actual state or with (b) the *adoption of a target* that is typically externally defined. In the former case, the person or group handling the issue will specify the goal to be achieved on their own based on the results of the analysis of the actual situation. In the latter case, the analysis of the gaps between the actual and the desired state will follow the target adoption activity.

In the next step, solutions for the now properly analyzed situation have to be conceived; the identified solution variants then have to be analyzed and compared. In design practice, this process is carried out as an alternation of steps for the *synthesis* of solutions and the *analysis* of their characteristics.

In the subsequent activity of *analysis and assessment*, the previously developed solution variants will be analyzed thoroughly; their achieved characteristics will be assessed and compared to the formulated target and with each other. As a result, a proposal for one or multiple solution variants is developed. Within the next step, the *decision* has to be taken whether one of the solution variants may be accepted as the basis of the further proceeding or whether the cycle has to be re-started with a refinement of the objectives.

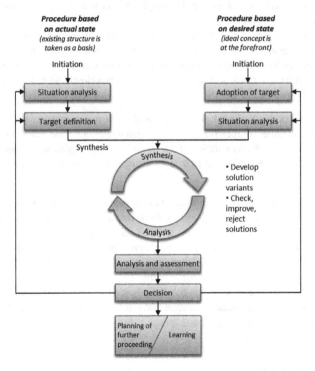

Figure 2.9: Cycle of problem solving at the micro-level, according to (Daenzer and Huber 1994; cited in VDI 2004)

The results obtained from one micro-cycle will form the input for subsequent cycles focusing on different problems. The content and sequence of subsequent cycles is conceived in the final step of *planning of the further proceeding*. Moreover, the end of each problem solving cycle should be considered as an appropriate moment for critical reflections on experiences gained and potentials for improvement.

2.4.2.1.2 Macro-Cycle: Procedure Model and Pre-defined Process Modules

The V-model originates in software engineering, where it belongs to the family of sequential procedure models and presents an extension to the well-known waterfall model. In general, these sequential development methodologies are characterized by a sequence of different phases: leading from requirements, through an architecture and design phase, which is followed by the actual implementation, to a final step of integration, verification, and testing of the different components and finally ending at the delivery of the complete software solution.

Gausemeier and Möhringer (2003) mention three reasons for the adoption of this procedure model within MPD:

(a) The model reflects well the top-down style of system design and the bottom-up approach of the system integration phases.
(b) It emphasizes the need for continuous validation of achieved properties of the system under design against the desired characteristics described by the requirements.
(c) Because the V-model based approach has already been applied in systems engineering and in MPD, a higher industrial acceptance for such a closely related procedure model may be expected.

Figure 2.10 depicts the V-shaped procedure model for a macro-cycle as well as the pre-defined process modules for recurring activities within one macro-cycle. Each iteration of the macro-cycle starts with the formulation of the *requirements* specifying the properties desired for the product to be developed and later to be assured throughout the subsequent development phases.

In the domain-spanning *system design* phase, these desired characteristics of the to-be product are translated into several main functions that divide into a hierarchy of sub-functions, the so-called *function structure*. For each of the identified sub-functions, appropriate Wirk- principles[16] and solution elements belonging to a particular domain have to be identified. This activity is called *"functional partioning"* and eventually leads to the Wirk-structure[17] (Jansen 2006). In a next step, the activity of *"spatial partioning"*, i.e. the positioning and grouping of the components of the system, results in the component structure[18]. The system design phase results in a domain-spanning solution concept that comprises the overall physical and logical operating characteristics and the types and structure of the product's components.

---

[16] In German: „Wirkprinzipien"

[17] In German: „Wirkstruktur"

[18] In German: „Baustruktur"

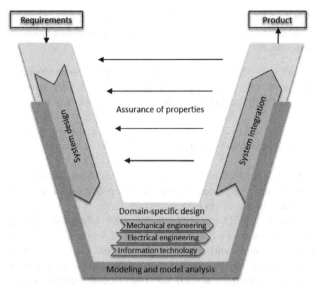

Figure 2.10: V-model-based procedure model for the macro-cycle, according to VDI guideline 2206 (2004)

Throughout the different development phases, several aspects of the system's behavior (e.g. dynamic, thermal, electric, magnetic) are examined by *modeling and analysis* using various types of models and tools. The domain-specific parts of the previously developed overall solution concept are then worked out concurrently in detail in the *domain-specific design* phase based on specific methodologies and tools.

Within the *system integration* phase, the domain-specific designs, i.e. the conceived subsystems and components, need then to be integrated into the context of the overall system. Jansen (2006) perceives the aforementioned activity of spatial partioning as closely related to the system integration phase because the component structure, i.e. the spatial partitioning's primary result, directs the system integration in combination with the disciplinary structures resulting from the domain-specific design activities. The higher the desired degree of integration, the more effort has to be foreseen for the checking of possibly undesired interactions between components originating in different disciplines.

Throughout the entire development process, the achieved quality of design has to be verified and assured. Here, the guideline foresees the continuous *assurance of properties*, i.e. the assessment of the achieved product characteristics – mainly based on modeling and model analysis, and the comparison of the current results with the desired characteristics.

Each macro-cycle results in a *product* with a certain degree of maturity determined by the quality of the coverage of the desired product characteristics. Examples for the different degrees of product maturity are e.g. prototypes at the levels of laboratory and function, and pre-series products.

2.4.2.1.3 Iterative Approach

Bellalouna (2009) emphasizes that iterations within the V-model-based procedure model may take place at four different levels:

(a) *Component* level: The characteristics of individual components will have to be improved in an iterative development process comprising multiple iterations of the component until finally leading to an optimized component fully covering the requirements.

(b) Level of *domain-specific system*: Similarly, the domain-specific sub-systems will be further developed until a configuration with sufficient requirements coverage has been reached.

(c) *System* level: At the level of the overall system, the different domain-specific systems will be integrated. This process will have to be repeated until a successful integration has been achieved that covers the requirements.

(d) Level of *macro-cycle*: Depending on the complexity of the mechatronic product to be designed, multiple macro-cycles may be required for conceiving the product with a sufficient degree of maturity. Usually, from one product iteration to the next, a higher degree of maturity may be achieved. The number and actual content of each of the macro-cycles may be freely adapted according to the targeted product maturity and the overall product complexity.

2.4.2.1.4 User-Specific Process Modules

In addition to the usage of the previously described pre-defined process modules, the guideline foresees the flexibility to adapt these existing process steps or to add new ones according to the situational conditions for the specific product or company (Gausemeier and Möhringer 2003).

2.4.2.1.5 Conclusions

The VDI guideline 2206 introduces an innovative approach for the interdisciplinary development in MPD as it integrates (a) elements of iterative development (e.g. the variation of the number of macro-cycles) with (b) practices of simultaneous engineering (e.g. the parallel activities within domain-specific design and the recommendation for interdisciplinary project teams) into its integrative but (c) mainly sequential process model. Although it currently provides the most extensive procedure model for MPD (Felgen 2007), it still does not reach the level of a ready-to-use development methodology and should rather be considered as a framework concept (Vajna, Weber, et al. 2009).

Chami et al. (2010) dismiss the V-model-based procedure as the definite solution for MPD and perceive it as a theoretical construct without the necessary tool support. In particular, they criticize that:

(a) the V-model inherits the unfavorable long cycle times from the aforementioned sequential progress model, and that

(b) the process follows a top-down approach, where the system design phase results in the domain-spanning solution concept that is later concretized in the involved domains. Only during the domain-specific design phase, however, new knowledge will be acquired that challenges parts of the previously established solution concept. The subsequent adaptation of the solution concept will require another time-consuming iteration of the whole process.

Jansen (2006) focuses his critical comments on the lack of procedures, models and tools for the activities of functional and spatial partitioning within the guideline. Likewise, Stetter et al. (2011) note that the guideline does not propose a method or tool for establishing the solution concept.

Felgen (2007) criticizes the guideline for only recommending the existing domain-specific development methodologies for the domain-specific design, while it does not emphasize the relevance of the cross-domain communication during this phase, despite this being previously described by Eversheim et al. (2000) by the usage of so-called *"synchronization points"*. Likewise, Qamar et al. (2011) state that the guidelines does not address the management of the dependencies between the involved domains.

The present thesis has a particular interest in the procedure models for MPD as it aims at analyzing the knowledge characteristics of MPD (knowledge bases, structure of knowledge and knowledge flows). This attempt requires a clear understanding of (a) the different roles within the development process and their associated knowledge bases, (b) the performed activities using specific methods and tools, and (c) the associated flows of information and knowledge. Here, the VDI guideline 2206 offers an initial orientation; nevertheless, for all of the three aspects, more detailed information needs to be collected in addition to that provided by the guideline.

2.5 Mechatronics Engineering Science[19]

In the middle of the 1990s, the appearance of the first Mechatronics-oriented journals, e.g. *"IEEE/ASME Transactions on Mechatronics"*, marked a change in the status of Mechatronics, at that time gradually becoming an academic discipline (Pop and Mătieş 2009). Today, Mechatronics is taught at numerous universities and polytechnics in Germany and other countries underlining its status as a well-recognized scientific branch within mechanical engineering. Corresponding to its interdisciplinary content on the technological side, as an academic discipline it is widely regarded as an interdisciplinary field (Schweitzer 1989; Buur 1990; Wikander, Törngren, et al. 2001; Smaili and Mrad 2008; Alciatore and Histand 2011). The term *"interdiscipline"* stands for such a scientific field that starts in-between the bodies of knowledge of established disciplines and later on becomes an academic discipline on its own (Repko 2008). In addition to Mechatronics, other prominent interdisciplines include e.g. biochemistry, bioinformatics, biophysics, and systems engineering.

---

[19] Möhringer and Stetter (2010) contrast the term *Mechatronic Design* with the term *Mechatronics Engineering*: They perceive *Mechatronic Design* as a specialization of engineering design focusing on Mechatronics-specific aspects whereas *Mechatronics Engineering* combines the contributions of (a) systems engineering within MPD with the ones of (b) mechatronic design explicitly including physical aspects of the conceived products. The present thesis uses the latter designation as the super ordinate concept.

Besides this interdisciplinary perspective on Mechatronics, two other perceptions on its disciplinary character were found in the analyzed research literature: Firstly, Tomizuka (2002) does not consider Mechatronics as an interdisciplinary area at all. Instead, he views Mechatronics as a principle and *"'best practice' for synthesis by mechanical engineers and those in other engineering disciplines."* Although Tomizuka does not explicitly describe the academic structures for Mechatronics he foresees, it is obvious that he considers Mechatronics as a design paradigm influencing many other disciplines, and in this way transgressing the disciplinary boundaries. Secondly, Pop and Mătieş (2009; 2011) describe a related perspective on Mechatronics. They perceive Mechatronics as a transdisciplinary approach to engineering that (a) crosses the traditional boundaries of a discipline, and (b) permits the emergence of new thematic mechatronical disciplines, e.g. optomechatronics, robotics, biomechatronics (Pop and Mătieş 2009; Pop and Mătieş 2011).

Surely, the discussion on the disciplinary character of Mechatronics may be considered as a discussion of minor relevance. Concerning the present thesis, however, the knowledge bases and structures of Mechatronics engineering science have to be well understood because Mechatronics as engineering science acquires and supplies large parts of the knowledge applied within MPD and contributes therefore substantially to MPD's knowledge characteristics. For this reason, the existing concepts of disciplinary interaction will be outlined in a first step. Then, the different steps of the evolution of Mechatronics engineering science as academic discipline are presented and mapped onto the different disciplinary interaction models. Finally, the previous findings will be compiled in a concluding characterization of Mechatronics engineering science.

2.5.1 Disciplinary Interaction and Evolutionary Models

In the analyzed literature on MPD (cf. section 1.1), the different attributes describing the disciplinary interaction models, e.g. *"multidisciplinary"*, *"interdisciplinary"*, *"transdisciplinary"* were typically used (a) without referring to clear definitions, and (b) the meanings of these attributes differed significantly between the publications (Neumann 2012). Moreover, even the literature on interdisciplinarity research does not use this terminology consistently. In order to provide the necessary clarifications, this section analyzes and utilizes the interpretations provided by two "de facto" reference publications on interdisciplinarity research (Klein 1990; Frodeman, Klein, et al. 2010). As of 2012, Klein's book *"Interdisciplinarity: History, theory, and practice"* ranks as the most cited book on interdisciplinarity, whereas the *"The Oxford Handbook of Interdisciplinarity"*, which appeared in 2010, has the potential to become the standard reference work in this research domain in the near future.

Klein (1990) states in her book on *"Interdisciplinarity: History, theory, and practice"* that there is consensus in the scientific community only on the two distinctions (a) multidisciplinarity vs. interdisciplinarity, and (b) interdisciplinarity vs. transdisciplinarity. She accordingly uses these distinctions to describe the terms by contrasting them with each other.

Multidisciplinarity is described by her as a *"juxtaposition of disciplines"* in a strictly additive manner without any explicit cooperation between the different disciplines (Klein 1990). The relationships between the disciplines remain limited and temporary, and no exchange of scientific methods and procedures will take place (cf. Figure 2.11). Therefore, the existing structure of knowledge of the involved disciplines is not questioned, and the disciplines remain unaltered (Klein 1990). Each of the involved disciplines handles its research subject in its specific way (Brand 2004), although these disciplinary research subjects relate to the same overall research problem. Furthermore, the research subjects of the involved disciplines may partially overlap. Finally, the different disciplinary results will simply be added, but not merged to an integrated result (Zirkler 2010).

According to Frodeman et al. (2010), the line between multidisciplinarity and interdisciplinarity will be crossed once *"integration and interaction become proactive"*, as depicted in Figure 2.11 by the arrow marking the transition between multidisciplinary activities (*Phase I*) and interdisciplinary interactions (*Phase II*). Furthermore, *interdisciplinarity* is characterized as an integrative approach (Klein 1990), where multiple disciplines conduct research on a common subject (Brand 2004). Frodeman et al. (2010) emphasize that (a) through interdisciplinary interactions the issues and questions common to the involved disciplines will be linked, and that (b) the existing disciplinary approaches will be restructured through explicit focusing and blending (*Phase II*).

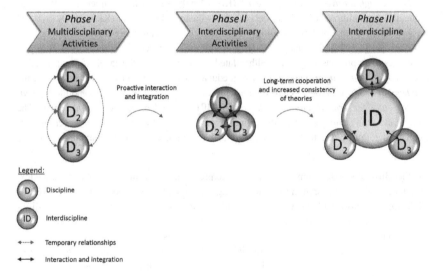

Figure 2.11: Evolutionary model reaching from multidisciplinary interactions to the emergence of an interdiscipline

Klein (1990) describes four kinds of interactions that have the potential to induce an interdisciplinary approach:

(a) Through *borrowing,* one discipline adopts concepts and/or methods from other disciplines either for a limited period of time or as part of a long-term relationship between these disciplines.

(b) The process of *solving of problems* may induce a cooperation of several disciplines even though the participating disciplines do not intent to integrate conceptually their bodies of knowledge.

(c) When two - typically neighboring - disciplines work in the same research area over a long period, they may reach a higher level of *consistency of subjects or methods* in the borderline area of the two disciplines.

(d) Finally, a long-term cooperation of several disciplines in connection with an increased consistency of their theories for a specific research area (indicated by Figure 2.11 as transition to *Phase III*), may finally lead to the *emergence of an interdiscipline* that exhibits its own body of knowledge as indicated in Figure 2.11 by the newly appearing interdiscipline *ID*.

Transdisciplinarity constitutes a breach to the multidisciplinary and interdisciplinary interaction modes that preserve the traditional mode of knowledge production within the disciplinary organizational model of science. Transdisciplinarity, however, aims at a different mode of gaining knowledge beyond the limits of disciplines (Frodeman, Klein, et al. 2010). It is characterized as a common system of paradigms that transcends the limits of traditional disciplines *"through an overarching synthesis"* (Klein 1990; Frodeman, Klein, et al. 2010). Consequently, the traditional disciplines become subordinate and irrelevant (Klein 1990). Systems theory, structuralism, and sociobiology are often cited examples for transdisciplinary approaches (Klein 1990). Nowadays, the transdisciplinary approach is actively followed in particular research projects (e.g. for health related problems under the label of *"transcendent interdisciplinary research"*) and promoted in education and research by organizations like the *Academy of Transdisciplinary Learning and Advanced Studies* (The ATLAS) and the *International Center for Transdisciplinary Research* (CIRET) (Frodeman, Klein, et al. 2010). The current and future impacts of transdisciplinarity on the scientific landscape, however, are difficult to determine, due to the far-reaching and discontinuous-revolutionary nature of this concept.

Another difficulty arises from the deviating interpretation of transdisciplinarity in the context of interdisciplinary research in Germany, as e.g. described in the proceedings of the THESIS workshop on transdisciplinarity (Brand, Schaller, et al. 2004). In this context, the transdisciplinary interaction model has been characterized by the notions that (a) the involved disciplines establish mutual relationships, and that (b) they transcend their disciplinary borders in order to move closer to neighboring disciplines (Brand 2004). The continuous stretching of the disciplinary boundaries in combination with the intense transdisciplinary collaboration may eventually lead to a situation where the definition of a discipline based on its boundaries becomes impossible (Brand 2004). From the original disciplines involved, a new discipline may then emerge that focuses entirely on the particular research subject (Brand 2004), as in the case of biochemistry or Mechatronics. The transdisciplinary interaction and evolutionary model as laid-out by Brand (2004) conflicts in the following two points with the previously described perception of interdisciplines and transdisciplinarity (Klein 1990; Frodeman, Klein, et al. 2010):

(a) The emergence of new disciplines through transdisciplinary interaction contradicts the previously described overall aim of transdisciplinarity to gain knowledge beyond the limits of disciplines and beyond the disciplinary organizational model of science.

(b) It also contradicts the described evolutionary theory for the emergence of interdisciplines, which locates the process of the emergence of new disciplines in the interdisciplinary context.

As a result of these inconsistencies of the model of transdisciplinarity described by Brand (2004) with the theories given by Klein (1990) and by Frodeman et al. (2010), the present thesis decides to follow the coherent interpretations of transdisciplinarity given by Klein (1990) and by Frodeman et al. (2010). These theories allow a consistent picture to be drawn for each of the three disciplinary interaction models and their associated evolution (cf. Figure 2.11). On this basis, Table 2.1 summarizes the identified characteristics, the pursued organizational model of science, and the relations between the disciplines and the research subject for each of the disciplinary interaction models. Due to the qualitatively different nature of transdisciplinarity, the table sets transdisciplinarity visually apart from the two other concepts of disciplinary interactions.

Table 2.1: Characteristics, science structure and relations to research subject of disciplinary interaction models

	Multidisciplinarity	Interdisciplinarity	Transdisciplinarity
Characteristics	• Juxtaposition of disciplines • No explicit cooperation, only temporary and limited relationships between disciplines • Disciplines and their knowledge structure not altered	• Proactive interaction and integration between disciplines • Linking of issues and questions common to involved disciplines • Restructuring of existing disciplinary approaches	• Common system of paradigms • Transcending the limits of traditional disciplines
Science Structure	• Within the limits of the disciplinary organizational model	• Within the limits of the disciplinary organizational model	• Beyond the disciplinary organizational model
Relations to Research Subject	• Focus on discipline-specific research problems, related to the overall research subject • Partial overlapping of discipline-specific research problems	• Common research subject for all disciplines	• Holistic view on research subject

2.5.2 Disciplinary Interactions and Evolution of MPD and Mechatronics Engineering Science

Starting from the findings on disciplinary interaction modes outlined in the previous section, the present section uses these theoretical foundations to supply answers to the first of the research questions formulated in section 1.2:

What is the most appropriate disciplinary interaction model describing the nature of the cross-disciplinary collaboration in MPD and in Mechatronics engineering science?

In the case of MPD, the evolutionary model of mechatronic products (cf. Figure 2.4), as outlined in section 2.2, provides clear indications which types of disciplinary interactions are encountered in the different evolutionary phases of mechatronic products. For the first kind of mechatronic systems (*Phase III* = functionally integrated mechatronic systems) the evolutionary model provides an indication that the interaction between the disciplines remains on the multidisciplinary level. Using a subsystem based approach, first the interfaces between the technologically homogenous components are designed, and in the next step, each discipline develops its subsystems on its own without a need for extensive communication and integration (Wikander, Törngren, et al. 2001). The interaction style during this phase of functional integration may be characterized as *multidisciplinary*, because the described approach does not require a *"proactive interaction and integration"*.

The subsequent phase (*Phase IV*) focuses on the spatial integration of the already functionally integrated components. It requires a closer collaboration between the disciplines involved to address the challenging mechanical, thermo-dynamic and spatial conditions for the spatial integration of the heterogeneous components (Zohm 2003). The different disciplines have to interact extensively throughout all phases of product development, which leads to the conclusion that the line to *interdisciplinary* interaction is crossed. Among the four types of interactions inducing an interdisciplinary approach (Klein 1990), the described interaction scenario for spatial integration belongs to category of *"solving of problems"*. All of this equally applies to the fifth development stage of mechatronic products (*Phase V*) that focuses on the further optimization of the already functionally and spatially integrated system.

In the following, the nature of the disciplinary interactions in the different evolutionary phases of Mechatronics engineering education and research will be investigated based on the previously introduced evolutionary model for disciplinary interactions (cf. section 2.5.1).

Grimheden and Hanson (2005) studied the evolution of Mechatronics in engineering education and propose a six-stage model for the development of Mechatronics as an academic discipline (cf. Figure 2.12). After an initial stage without any disciplinary interactions (stage 1), a multidisciplinary phase (stage 2) follows where students, who wish to broaden their knowledge in neighboring disciplines, simply attend the existing, not adapted courses offered by those departments. In the third phase, the different departments agree on necessary adaptations of the existing courses toward the specific needs of Mechatronics education.

Figure 2.12: Model by Grimheden and Hanson (2005) for the evolution of Mechatronics as academic discipline

In the fourth stage, described as the curriculum stage, the involved departments specify new curricula for Mechatronics. In the fifth stage, new academic structures, e.g. a Mechatronics department, emerge that assume the organizational responsibility for the offered courses in this field. This development finally leads to a situation where Mechatronics is considered as a discipline with its own *thematic identity* (Grimheden and Hanson 2005).

Although the Grimheden and Hanson (2005) model and the previously introduced evolutionary model for disciplinary interactions (cf. section 2.5.1) coincide in the overall development directions toward a dedicated identity of Mechatronics as academic discipline, Grimheden and Hanson (2005) do not explicitly label the interactions among the engineering disciplines offering courses relevant to Mechatronics as interdisciplinary. Their descriptions of the disciplinary interactions leading to the emergence of the new discipline, however, indicate that the line toward *interdisciplinary* activities has been crossed. Likewise, their classification of Mechatronics as *thematic* discipline conforms to the characterization of Mechatronics as interdiscipline for the reason that an interdiscipline emerges from activities on a *common topic* or *theme*.

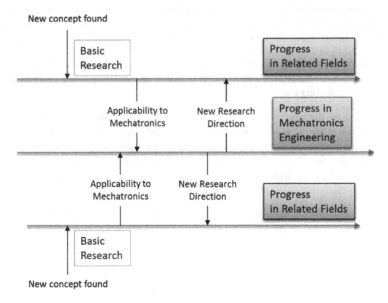

Figure 2.13: Research interactions of Mechatronics engineering with related engineering disciplines, according to (Wikander, Törngren, et al. 2001)

Beyond the educational aspects, Wikander et al. (2001) analyzed the *research* interactions of Mechatronics science with related engineering disciplines. They perceive Mechatronics engineering science in a dual role to (a) validate and integrate research results from related disciplines, and to (b) identify and conduct new research directions that lead to new interdisciplinary research activities (Wikander, Törngren, et al. 2001).

Within Mechatronics interdisciplinary research, Mechatronics science has its strength in providing the applicative and integrative frame whereas the related engineering fields contribute their research results and benefit from the new or extended research questions raised by Mechatronics (cf. Figure 2.13).

2.5.3 Conclusions

In the course of answering the first of the research questions, the previous section showed that interdisciplinarity is the prevailing disciplinary interaction model encountered both in MPD and in Mechatronics engineering science. Typically, the disciplinary interaction starts with a phase of multidisciplinary activities characterized by a loose and temporary cooperation followed by an interdisciplinary phase where the disciplines proactively interact and integrate their bodies of knowledge within a common field of interest. Eventually, Mechatronics emerged as an interdiscipline from mechanical engineering, electrical engineering, computer science, and control engineering through the long-term interdisciplinary fusion of their theories, concepts, methods, and tools within the area of the development of heterogeneous technical systems. Following the interpretations of transdisciplinarity given by Klein (1990) and by Frodeman et al. (2010), transdisciplinarity can be excluded as explanatory model for the disciplinary interactions in Mechatronics.

For the targeted analysis of the knowledge characteristics of mechatronic product development in the frame of the present thesis, the confirmed interdisciplinary nature of interactions implicates the following issues:

(a) Due to the described character of Mechatronics as interdiscipline, a knowledge level analysis of MPD has to take into account individually the characteristics of (i) Mechatronics engineering and (ii) each of the other disciplines involved in MPD.
(b) Likewise, for an analysis of the knowledge flows between the different disciplines and/or roles collaborating in MPD, Mechatronics engineering and the associated roles of e.g. the system architect have to be separately taken into account.

Figure 2.14 summarizes the insights gained in the last three sections on the disciplinary structure, the characteristics in the research area and in education of Mechatronics engineering science.

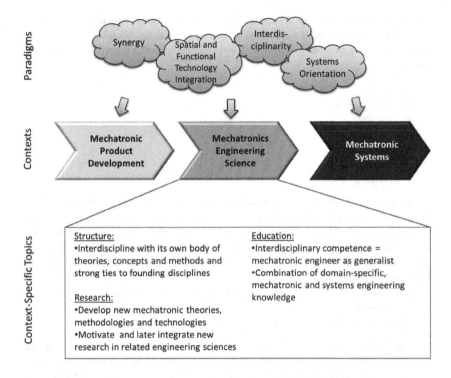

Figure 2.14: Manifestations of the overall mechatronic paradigms in the context of Mechatronics engineering science

2.6 Modeling and CAx Tools in Mechatronic Product Development

Within multi-disciplinary product development, the number of digital models generated and exploited for each component may easily exceed the mark of one hundred (Stark and Damerau 2011). This illustrates the vital role of models and in a wider sense of model-based development within multi-disciplinary design. Model-based development enables among others the realization of virtual prototyping as it provides the models as well as associated methods contributing to the virtual prototypes. In combining them, they may be perceived as *"a way of buying information of the final product, and thereby minimizing the risk of making false decisions"* (Buur and Andreasen 1989). The associated process of acquiring knowledge for a product under development starts with activities of learning, followed by a phase of the application of the gained knowledge through experimenting (Avgoustinov 2007). Within this process of experimentation, the use of virtual models instead of physical ones has the potential (a) to make the extensive experimentation affordable, and (b) to increase the efficiency of learning (Avgoustinov 2007).

In a broader context, Buur and Andreasen (1989) propose a morphology of design modeling that perceives the purposes of modeling as being identical with the following basic operations of engineering design, which the designers apply repeatedly during the product development process:

(a) Define
(b) Generate ideas
(c) Describe
(d) Verify
(e) Evaluate
(f) Specify
(g) Arrange information

Overall, their morphology of design modeling (cf. Figure 2.15) comprises the description of (a) the modeling activity and of (b) the resulting design model. The modeling activity primarily focuses on (i) the original object to be modeled[20], (ii) its relevant properties, i.e. the attributes and functions incorporated in the model (cf. Figure 2.16), (iii) the aforementioned modeling purpose, and (iv) the intended types of users for the model that largely determine the modeling language to be applied. Buur and Andreasen (1989) consider the design model as a means for communication within the design process that is consequently characterized by (i) the specifics of the applied language, (ii) the characteristics of the medium used for the model, (iii) the level of abstraction, and (iv) the level of detail (complexity).

As an application example, the previously described process of experimentation may be characterized in terms of this design morphology by the activities of (a) *verification* and *evaluation* of the already gained knowledge on particular aspects of the product under development, and of (b) *generation* of new insights and ideas.

---

[20] Avgoustinov (2007) proposes the term *modellee* to represent the object of modeling.

The Modeling Activity			
Object	Property	Purpose	User
Problem	Design	Define	The designer himself
Criteria	Time consumption	Generate ideas	Colleague
Process structure	Costs	Describe	Project team
Technological	...	Verify	Draftsman
principle	Manufacturing	Evaluate	User
Order of operations	Tolerances	Specify	Salesman
...	Assembly time	Arrange information	Client Manager
Function structure	...		...
Necessary effects	Marketing		Computer
Interface	Design appeal		NC machine
...	Ease of packing		...
Organ structure	...		
Working principles	Usage		
...	Function		
Parts structure	Ease of operation		
Total form	Reliability		
Dimensions	...		
...	Destruction		
	Impact on		
	environment		
	...		

The Design Model	
Code	Medium
Language	Talking
Symbols	Graphics
Letters	Paper
Numbers	Photograph
Mathemati-cal terms	Movie, video
Electrical symbols	Computer display
Mechanical symbols	...
Drafting symbols	3 Dimensional
Flow chart symbols	Standard components
...	Raw materials (paper, cardboard,
Reproduction	wood, plastic, foam, metal, clay,...)
Projected	...
Spatial	
...	
Others...	

Concrete ↓ Level of abstraction → Abstract

Detailed ↓ Number of details → Simple

Fine ↓ Manufacturing technique → Coarse

Figure 2.15: Morphology for design modeling, adapted from (Buur and Andreasen 1989)

Likewise, Avgoustinov (2007) analyzed the rationale for the application of models and identified more detailed purposes for modeling, which are presented in the following with an associated reference to the purposes within the design morphology of Buur and Andreasen (if applicable):

(a) Description of an idea (*Define* and *Describe*)
(b) Common basis for discussions and information exchange (*Describe, Specify* and *Arrange information*)
(c) Simplified representation of the modeled object that comprises only the characteristics of interest for the particular activity, as depicted in Figure 2.16. This reduced complexity allows for
    i.   an easier comparison of the models as substitutes of the modeled objects, e.g. the modeled components or complete products (*Verify* and *Evaluate*),
    ii.   an easier handling of the model with respects to the potentially complex structure of the modeled object (relevant for *all* operations), and
    iii. an easier understanding of the complex matter transformed into the model (relevant for *all* operations).
(d) Prediction of the behavior and characteristics of the modeled object, to foresee problems and weaknesses (*Generate ideas, Verify* and *Evaluate*)
(e) Possibility to analyze non-existent or not available objects (*Verify* and *Evaluate*)

Whereas the design morphology of Buur and Andreasen includes physical and virtual models essentially on equal terms, the focus in the current practice of MPD shifts to virtual models. For the building and exploration of these virtual models, many kinds of software platforms are applied throughout the product development process – these are usually subsumed under the umbrella term *"CAx tool"* (e.g. CAD, CAE, EDA, function modeling tools, mathematical modeling tools, system-level modeling tools).

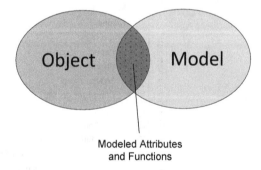

Modeled Attributes
and Functions

Figure 2.16: Model as a representation of selected attributes and functions of the original object, after (Buur and Andreasen 1989)

Andreasen (1994) summarizes the central role of modeling within product development underlined by the presented numerous purposes of modeling by stating that *"modeling is the language by which the designer or product developer can elaborate, synthesize, evaluate and communicate"*. In this spirit, the present thesis perceives the various types of models in MPD primarily as carriers of information and knowledge that are exchanged in widespread and complex networks in the course of the product development process. A deeper understanding of the characteristics and applications of models in MPD is therefore required for the analysis of the knowledge characteristics of MPD and the subsequent development of a design support system based on the semantic enrichment of models (cf. laid section 1.2). These topics will be addressed in the remainder of this section under the following headings:

(a) Definition of a modeling taxonomy for MPD
(b) Identification of modeling tools for MPD
(c) Allocation of models to design phases of VDI guideline 2206

2.6.1 Definition of a Modeling Taxonomy

In view of the high number of different types of models generated and exploited during the different design phases of MPD, a systematization of the practiced modeling techniques appears indispensable. Avgoustinov (2007) proposed a large set of about 40 classification criteria applicable to modeling of which he selected the following group of key criteria:

(a) *Application domain or sub-domain*: Within the engineering domain, modeling may target design, manufacturing, assembly, etc.
(b) *Medium of the model*: Virtual models vs. physical ones
(c) *Usage-related attributes*: Durability, interoperability, reusability, etc.
(d) *Structure-related attributes*: Relations between components, divisibility, etc.
(e) *Implementation-related attributes*: Behavior (static vs. dynamic), definiteness (deterministic vs. non-deterministic), etc.

Given that the set of classification criteria does not bear any special focus on MPD, it seems necessary to further specialize or, alternatively, restrict some of the aforementioned classification criteria. In the first step, selected classification criteria will therefore be adapted to take into consideration the specific understanding of MPD developed in previous sections:

(a) Section 2.1.2 outlines the engineering domains relevant within Mechatronics, which allows restricting the *application sub-domains* to mechanical engineering, electrical engineering, and computer science with control engineering as one of its sub-disciplines, as well as systems engineering due to the systems oriented nature of MPD discussed in section 2.4.1.
(b) Because of the strong adherence of MPD to model-based development and virtual prototyping combined with the focusing of these approaches solely on software models, the *medium of the model* as classification criteria may be omitted within MPD.

In the second step, additional criteria will be included covering (a) areas of special importance for MPD, (b) model traits potentially needed by the analysis of knowledge characteristics, and (c) specific properties possibly required for the intended semantic enrichment that have not been covered by the key criteria introduced by Avgoustinov (2007). In order to evaluate a modeling techniques' ability to reach beyond disciplinary borders, the taxonomy should include the domain-spanning capabilities (*vertical scope*) as criteria.

Furthermore, a classification of the modeling approaches according to their presence within the different design phases and pre-defined activities of a macro-cycle as defined by the VDI guideline 2206 (*horizontal scope*) will be helpful for the analysis of knowledge characteristics of MPD because it identifies the range of activities a particular type of model is involved in.

Next, additional criteria are required to describe the actual attributes and functions covered by the model, visualized by the intersecting area between the original object and its model in Figure 2.16. In the context of the analysis of knowledge characteristics of MPD, these criteria will permit a description appropriate to a model's content of information and allow comparison of models by their content. In the context of the intended semantic enrichment, these criteria indicate the model's properties, which are possibly captured by this process.

All together, the aforementioned criteria that characterize the reach of the particular modeling technique with regards to e.g. the covered disciplines and design phases, the included components and their relations, the covered artifacts related to the design process like e.g. requirements, functions, may be subsumed under the term *modeling scope*.

In addition, criteria classifying the applied modeling approach (e.g. procedural, declarative, object-oriented) need to be included. Essentially, the used modeling paradigm comprises both a procedure model and a specific meta-model at the data level to be applied during the modeling process. In this way, the modeling paradigm largely determines (a) the model's inner properties and structure and (b) its capabilities for obtaining specific sets of analysis results. The result data will be needed for both the analysis of knowledge characteristics of MPD and the intended semantic enrichment of the various models used in MPD.

As a whole, the following set of criteria will permit the establishment of a modeling taxonomy for MPD:

    (a) *Modeling scope*: Describes which of the product's constituents the model encompasses:

        i.   *Element scope*: Specifies which categories of system elements (e.g. requirements, functions, logical, physical) belonging to the overall product are taken into account by the model.

       ii.   *Relations scope*: Indicates the types of relations (e.g. structural relations, parametric relations, dependencies, geometric constraints) between the previously mentioned elements taken into consideration by the model.

      iii.   *Design phase scope* (or *horizontal scope*): Indicates which design phases are covered by a particular type of model. This criteria is tightly coupled with the *level of abstraction* within the *model-related characteristics*, because each design phase requires an appropriate level of abstraction of the model in order to achieve the required level of detail of analysis (Sinha, Paredis, et al. 2001).

      iv.   *Domain scope* (or *vertical scope*): Describes the capabilities of the model to capture the characteristics of one (single-domain) or multiple domains (multi-domain), or to act on the system level (domain-neutral).

       v.   *Time-related behavior*: In system theory, the external behavior of a system is defined as the relationship between the time histories of its inputs and outputs

(Zeigler, Praehofer, et al. 2000). In addition to the common distinction of static and dynamic systems, Vajna et al. (2009) include two special cases of systems and accordingly propose the following classification of technical systems:

- *Static*: A static system is a memory-less system of which the outputs directly depend on the values of the inputs.

- *Quasi-static*: Quasi-static systems may be sufficiently described by a sequence of states of equilibrium systems; the systems dynamics within these infinitesimal time spans is neglected.

- *Steady-state*: Steady-state systems are characterized by a state of equilibrium, i.e. the incoming and outgoing flows of substance, energy and information are balanced and keep the system's content on substances, energy and information at a constant level (Vajna, Weber, et al. 2009).

- *Dynamic*: A dynamic system exhibits a memory effect that causes the outputs to depend also on past and future values of the inputs. Dynamic systems may be further classified according to (a) the continuity and (b) the discreteness of the state variables with regards to time:

    - *Continuous*: A continuous system (also: analog system or *Differential Equation System Specifications* (DESS)) shows a behavior that changes continuously over time. Its state variables are determined by continuous functions of time (Khosrow-Pour 2007).

    - *Discrete time*: A discrete system (also: digital system or *Discrete Time System Specifications* (DTSS)) shows a behavior observable through its state variables changing only at discrete points of time (Khosrow-Pour 2007).

    - *Discrete event*: Within a discrete event system (also: *Discrete Event System Specifications* (DEVS)) the state variables change only at discrete points of time due to sudden events that are not previously known (Vajna, Weber, et al. 2009).

The determinateness of the system behavior upon a sequence of events at its inputs should be included as another classification criterion:

- *Deterministic*: A deterministic system shows upon a sequence of events at its inputs each time the same series of values at its outputs.

- *Stochastic*: In a stochastic system, the output values show a non-deterministic behavior upon an identical sequence of values at its inputs.

(b) *Modeling-paradigm*: The modeling-paradigm depicts the dominant modeling technique applied. In case of hybrid-modeling approaches, several paradigms have to be combined to sufficiently classify the chosen approach. Sinha et al. (2001) refer to a few dichotomies of common modeling-paradigms, for instance they contrast (a) graph-based and language based approaches, (b) procedural and declarative, and (c) functional and object-oriented languages. In order to capture approaches common in MPD, other criteria will be added, finally leading to the following systematic:

i.   *Procedural*: In procedural modeling languages, the value of an output variable is calculated by the assignment of the resulting value of a function of input variables (Sinha, Paredis, et al. 2001). The causality of the resulting mathematical relations is therefore fixed and requires a strict evaluation sequence that cannot be altered once the mathematical system has been defined. For this reason, *what-if* scenarios where the value of an original input variable is analyzed depending on the values of the former output variable cannot be investigated. The Continuous System Simulation Language (CSSL) is an early example of a procedural modeling language.

ii. *Declarative* or *equation-based*: Declarative modeling languages permit the definition of acausal equations, i.e. the status of a variable as input or output is not predefined. Declarative modeling languages support therefore (a) the investigation of *what-if* scenarios and (b) the reuse of predefined components within contexts requiring a different causality. Popular examples of declarative modeling languages are VDHL-AMS and Modelica (Sinha, Paredis, et al. 2001).

iii. *Object-oriented*: An object-oriented modeling language employs the principles of object-orientation to define the attributes, methods, and interfaces of each component, where certain of these properties may be inherited from base components. This methodology results in (a) an encapsulation of the component's properties within the component itself, in (b) the accessibility of the component's properties only through its public interfaces, and in (c) excellent abilities for the reuse of components within the inheritance hierarchy. Today, the most extensive support for object-oriented principles may be found in Modelica (Sinha, Paredis, et al. 2001).

iv. *Graph-based modeling*:
   - *Bond-graph*: Bond-graphs allow modeling of physical systems consisting of multiple domains (e.g. mechanical, hydraulic, electrical, and thermal) due to their focus on energy interaction. In principle, a bond-graph is a labeled and directed graph, where each vertex stands either for a component or a particular type of junction both comprising ports, and where an edge stands for an energy connection between these ports (Broenink 1999). Bond-graphs may also contain parts of block-diagrams that require signal inputs and outputs in addition to the power ports (Broenink 1999). Today, the bond-graph paradigm is supported by modeling languages (e.g. Modelica), graphical modeling front-ends (e.g. CAMP-G) and directly included in modeling and simulation tools (e.g. 20-sim, SYMBOLS Shakti and MS1) (Borutzky 2011). Unfortunately, the topology of bond-graphs does not match the topology of the modeled physical system (Sinha, Paredis, et al. 2001).
   - *Linear graphs*: A linear graph may be perceived as an abstraction of a circuit diagram (Borutzky 2011). For this reason, the linear graph represents well the topology of a physical system (Sinha, Paredis, et al. 2001). Within a linear graph, an edge represents an energy flow in a

component, whereas a node symbolizes a component's terminal (Sinha, Paredis, et al. 2001). As bond-graphs, linear graphs are able to represent physical systems comprising multiple domains.

- *Block diagrams*: In a block diagram, a block represents a component or sub-systems that are accessible through their inputs and outputs. Lines symbolize the flows of material, energy, or information between the blocks. Typically, coded procedures describe the behavior of the blocks (Sinha, Paredis, et al. 2001), which leads to identical limitations as described for procedural languages.

v. *Geometrical modeling*:

- *2D modeling*: In the past, the primary aim of 2D modeling used to be the generation of technical drawings typically containing different views of the particular product. Within each of the views, a set of 2D primitives (e.g. point, line, polygon, curve, etc.) and associated operations (e.g. trim, combine, mirror, etc.) were applied to generate a projected visualization of the product including all required dimensions. Today, 2D modeling is widely applied in the sketching modules of CAD systems that define the base geometries for all kinds of 3D operations, e.g. extrusion, sweep, etc. (Vajna, Weber, et al. 2009).

- *3D modeling*: 3D models permit the capture of the spatial information of a product, which offers substantial benefits over 2D models. Vajna et al. (2009) mention significant advantages such as (a) the early identification of possible functional or assemblage problems, and (b) the ability to integrate CAE and CAM tools on the basis of the 3D-model. Today, solids are the dominant modeling type among CAD-tools. Typically, the solids are internally represented by Boundary Representation (B-rep). It describes the 3D shapes using a combination of (a) their topology (e.g. faces, edges, vertices) and (b) geometrical information (e.g. points, curves and surfaces). In combination, solid modeling backed by B-rep-based geometrical modelers constitute today's technological mainstream that allows for the following relevant modeling approaches:

  - *Direct* or *explicit modeling*: Using direct modeling, the user adds the elements comprising the 3D model explicitly. Typically, the history of modeling actions will not be editable. Direct modeling is one of the oldest modeling approaches in CAD and has been applied since the mid-1960s (Yares 2011). In recent years, direct modeling tools benefited from new capabilities for dynamic feature inference that (a) enabled the modification of models previously established in other CAD tools, and (b) made explicit modeling attractive and usable to people involved in the product development process with significantly lower CAD skills. Consequently, a new generation of direct modeling tools is now applied to prepare CAD models for CAM and CAE tools, and in conceptual design (Ushakov 2011).

- *Feature-based modeling*: In contrast to direct modeling, feature-based modeling permits working with features, i.e. building-blocks combining a dedicated semantics (e.g. form, functional, assembly) with geometrical elements and operations (Ovtcharova, Weber, et al. 1997; Vajna and Podehl 1998; Vajna, Weber, et al. 2009) whereas direct modeling focuses on lower level geometrical primitives. Each feature-based 3D model contains a history of features that declares the chain of actions to obtain the final model. Additionally, the feature list permits the modification of the resulting model through the editing of any of the involved features. Typically, a feature is controlled by different parameters that are either defined by values entered by the user or by functions of other parameters (e.g. formulas or procedures). Therefore, parametric modeling forms a basis for feature-based modeling.

Figure 2.17 depicts the previously introduced hierarchy of criteria that permit the establishment of a modeling taxonomy for MPD.

The modeling taxonomy derived from these criteria will be applied to classify the models generated, communicated, and analyzed in the course of the different stages and by the different disciplines of the mechatronic development process. Moreover, the modeling taxonomy aims to capture and abstract the specifics of each model within a modeling type characterized by the aforementioned criteria. Models belonging to the same modeling type do not necessarily comprise an identical content. Instead, they match in (a) their semantics concerning the types of their input and result data, and (b) the purposes the models might be applied for. Consequently, this classification approach leading to modeling types will permit the neglection of the association of the model with a particular tool and its specifics to the greatest extent.

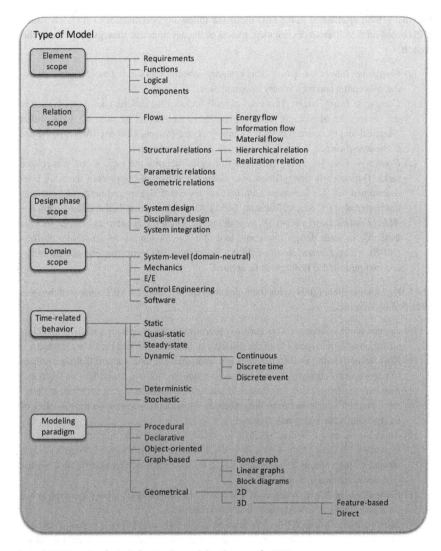

Figure 2.17: Hierarchy of criteria forming the modeling taxonomy for MPD

2.6.2 Identification of Modeling Tools

Within the development process of mechatronic systems, a high number of modeling tools and associated models are applied. In order to gain a sufficient overview on this subject, those modeling and analysis tools relevant in the different stages and disciplines of MPD will be identified by means of a literature search.

Kümmel (1999) established in his PhD thesis the following systematic for CAx tools within MPD based on the different development phases of the development strategy proposed in this context:

(a) *Computer-Aided Conception*: This category subsumes all kind of tools applied during the conceptual and preliminary design phases.

(b) *Computer-Aided Design*: This class of tools includes the different design tools used in the various domains, e.g. CAD tools applied in mechanical engineering, EDA within electrical and electronics design, *Computer-Aided Control Design* (CACD) tools, and *Computer-Aided Software Engineering* (CASE) environments.

(c) *Computer-Aided Analysis*: This group of tools supports the various analysis related tasks. During early design iterations, where the shapes of the product have not been determined yet, *Mathematical Software*, i.e. computer algebra and numerical methods environments (e.g. Maple, Mathcad, MATLAB, Mathematica, Scilab), and *Computer-Aided Control Design* (CACD) tools dominate. Once the geometrical shapes become available in later design iterations, tools for the *Simulation of Multibody Systems* (MBS), *Finite Element Analysis* (FEA), the analysis of electromagnetic compatibility, and computer-aided testing will be applied.

The VDI guideline 2006 (2004) adds the following relevant tools for MPD on top of the ones identified by Kümmel:

(a) Requirements elicitation: This class of tools aims at the systematic gathering and classification of requirements from the different stakeholders toward the future product.

(b) *Requirements management*: Requirements management tools support the management of requirements changes, the traceability of requirements, and check their consistency.

(c) *Function modeling* (also: Functional Modeling): Function modeling tools aim at deriving a product's functions from the gathered requirements, and support the modeling of the function's hierarchy and structure.

(d) *Boundary Element Method* (BEM): In contrast to the Finite Element Method, the Boundary Element Method requires a discretization of the analyzed object's surface instead of the volume. It is applied e.g. in fluid mechanics, thermodynamics, acoustics, and electromagnetics.

(e) *Computational Fluid Dynamics* (CFD): This class of methods allows the analyses of flow processes in thermodynamic and fluid mechanics.

(f) *Hardware-in-the-loop* (HIL): The idea behind a HIL simulation is to test physically available components (hardware) in a test environment that provides a real-time emulation for the missing parts of the product that are interfaced by the hardware to be tested.

Vajna et al. (2009) provide an exhaustive overview on CAx tools and related approaches for modeling, simulation, and optimization in product development without any special focus on Mechatronics. Besides the previously mentioned tools, they describe the following methods applied in MPD:

(a) *Co-simulation*: Co-simulation is a technique that addresses the need to combine and synchronize the simulation of the different aspects of heterogeneous systems typically executed by different tools.

(b) *Bond-graph modeling*: cf. section 2.6.1

(c) *Object-oriented modeling*: cf. section 2.6.1

(d) *Modeling and simulation of hydraulic systems*: Circuit diagrams for hydraulic systems describe the hydraulic components and their connections. The behavior of hydraulic systems is typically simulated in tools like MATLAB/Simulink, Modelica, or Automation Studio based on component libraries for hydraulic elements.

Likewise, the 2D layout of pneumatic systems is described based on circuit diagrams, whereas their behavior is simulated in tools like Modelica or Automation Studio providing pneumatic component libraries.

In recent years, several authors have emphasized the need for an appropriate tool support of system-level modeling activities during system design (Bellalouna 2009; Hehenberger, Egyed, et al. 2009; Chami, Seemüller, et al. 2010; Follmer, Hehenberger, et al. 2010; Qamar, Wikander, et al. 2011; Stetter, Seemüller, et al. 2011). The proposed ideas for the realization of system-level modeling span across a variety of approaches:

(a) Bellalouna (2009) proposes the realization of interdisciplinary modeling by means of the extension of a PDM system's data models and functionality.

(b) Follmer et al. (2010) describe the usage of the *Systems Modeling Language* (SysML) for the modeling of the requirements, structure, and behavior of mechatronic systems.

(c) Several authors (Chami, Seemüller, et al. 2010; Qamar, Wikander, et al. 2011; Stetter, Seemüller, et al. 2011) propose the application of the SysML model as the central instance within MPD for the integration of data from various disciplinary tools

Furthermore, first industrial realizations of integrated MPD environments have appeared supporting system-level modeling in combination with other capabilities:

(a) EPLAN Engineering Center (2004) (Steck 2008; Sendler 2009):
   i.   Component-based decomposition of mechatronic systems
   ii. Integration of disciplinary tools
   iii. Integration of libraries comprising generic and customer-specific components
   iv. Automatic generation of product documents, e.g. BOM, mechanical drawings, electrical schematics, and programs for PLC

(b) CATIA V6 Systems (2009) (Dassault Systèmes 2009; Dassault Systèmes 2011):
   i.   Multidisciplinary modeling and simulation of systems based on Modelica
   ii. Usage of Modelica standard component libraries
   iii. Definition of functional system architectures and breakdown into logical architecture
   iv. Modeling of control logic, code generation and simulation of behavior
   v.   Collaborative engineering between different roles and disciplines

vi. Requirements traceability spanning across the different phases of product development and product introduction based on the mapping of requirements on functions, logical elements of the product structure and test cases

(c) Siemens PLM Mechatronics Concept Designer (2010) (Siemens PLM 2010; Siemens PLM 2011):

    i. Conceptual design of mechatronic systems based on functional modeling

    ii. Physics-based simulation including kinematics, dynamics and collisions

    iii. Requirements management ensuring requirements traceability

    iv. Libraries consisting of objects with embedded knowledge

    v. Integration of disciplinary tools

    vi. Collaborative engineering between different disciplines

Based on the described research activities and the emerging tool support, it may be concluded that system level models are highly relevant within MPD and will therefore be considered when allocating types of model to design stages.

2.6.3 Allocation of Models to Design Phases of the VDI Guideline 2206

In the next step, the previously identified types of models will be allocated to the different phases of the macro-cycle from the VDI guideline 2206. In order to reduce the number of considered types of models to a range allowing handling the most relevant of them in detail, models with lower importance or with an optional character will be collected in the category *Other models*.

The *task clarification* phase generates a *requirements* list that specifies the properties desired for the product to be developed. This requirements list serves as input for the subsequent *system design* phase, where the requirements become part of the overall system-level models. In this context, additional requirements may be derived from the originally passed ones and links may be established to functions and other system elements implementing requirements. As a result, the requirements will emerge to a structure on their own as indicated in Figure 2.18.

Other system-level models relevant for the system design phase are the function structure, Wirk-structure, and component structure. In the context of embedded systems and automotive engineering, though, a rather electrics/electronics (E/E) minded approach with different terms and models has been established: Here, the *"logical system architecture"* comprises the content of the function structure with the addition of logical signals for the description of the communication between functions (Reif 2009). The *"technical system architecture"* allocates these functions to control units and maps the logical signals to the bus system connecting the control units (Reif 2009). In short, it assumes the combined roles of the Wirk-structure and the component structure.

Physics-based multi-domain models permit modeling and investigation of the behavior of a mechatronic system spanning over multiple domains early on during the system design phase.

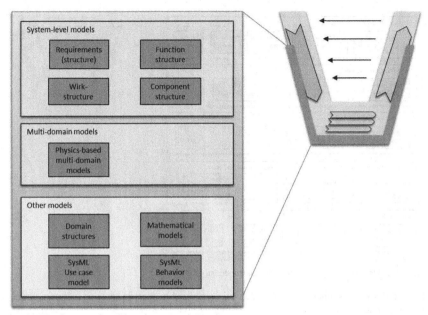

Figure 2.18: Allocation of models to the system design phase

In Figure 2.18 the category *Other models* contains models with a lower relevance or with an optional character for the system design stage: For instance, domain-specific structure models may be required when following an explicit partitioning approach, where each of the domain-neutral structure models (function, working, and component structure) is complemented with domain-specific equivalents supporting the explicit partitioning (Jansen 2006). Additionally, mathematical models may be applied for purposes of design analysis or synthesis during system design and later stages. Additional models defined within SysML may be an option for the description of the targeted use cases and the detailed specification of the behavior (activity diagram, sequence diagram, state machine diagram).

As a first step of the *disciplinary design phase*, discipline specific function structures need to be derived from the system-level function structure unless they were already supplied in case of explicit partitioning. In a next step, disciplinary system architectures will be developed based of the overall component structure. As depicted in Figure 2.19, each of the involved domains will then proceed to develop their specific design models (e.g. geometrical model, E/E layout, control design model, source code or formal model) and associated models for analysis and simulation (e.g. FEA, MBS, HiL, Software-in-the-Loop model (SiL), control simulation model). Alternatively, physics-based models may be applied for the simulation of problems placed in multiple domains.

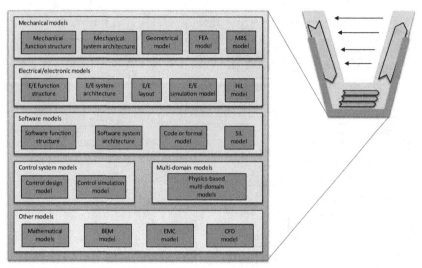

Figure 2.19: Identified models within the disciplinary design phase

In addition, various other types of models might be used in specific cases as indicated under *Other models*.

During the *system integration phase*, the components developed during the previous phase will be integrated into modules in a step-wise process depending on the chosen integration approach, e.g. modular integration, spatial integration, or integration of distributed components (VDI 2004). The component structure developed during the system design phase provides the necessary guidance for the integration process that is usually supported by means of digital mock-up (DMU) models (cf. Figure 2.20). DMU tools use the geometrical models originating from CAD to detect collisions between components and to simulate assembling/disassembling sequences. In addition to such geometry-oriented analysis steps, the system integration requires conducting complementary types of analysis for e.g. thermodynamic, EMC, or noise-related problems that may arise from the interaction of the multidisciplinary components. Due to the targeted level of detail and the availability of detailed component shapes, existing physics-based multi-domain modeling tools may not be able anymore to handle the analysis tasks during this stage. Therefore, the need for co-simulation environments arises that are capable of handling the simulation of coupled problems at a sufficient level of detail for system integration. The *FunctionalDMU* project is an example for such an initiative and tool aiming at (a) providing a comprehensive view on the functional aspects of the mechatronic product, and (b) the integration and execution of a multitude of simulation tools for the analysis of the system's behavior (Stork 2010).

Moreover, HiL and SiL-environments may be applied to work with combinations of physical and virtual components of the system under integration during simulation and testing. Finally, the approved configurations at each level of integration (modules, assemblies, and product)

have to be established and managed within each of the involved disciplines as indicated by Figure 2.20.

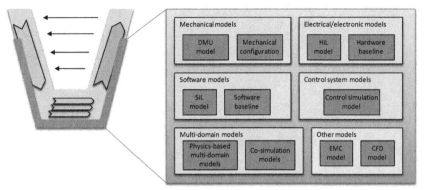

Figure 2.20: Relevant models for the system integration phase

So far, the relevant types of models relevant within the different phases of the macro-cycle from the VDI guideline 2206 have been identified. Each of the model types has been characterized further by the following two criteria originating from the previously established modeling taxonomy:

(a) Domain scope
(b) Design phase scope

Altogether, approximately 30 different types of models have been identified.

3 Knowledge: Concepts and Taxonomies

We have more information now than we can use, and less knowledge and understanding than we need. Indeed, we seem to collect information because we have the ability to do so, but we are so busy collecting it that we haven't devised a means of using it.

Warren Bennis, 1997

In the above quote, Bennis (1997) introduces one of the key challenges of information science that consists of the transformation of abundantly collected information into scarce knowledge that possesses the unique trait to enable actions. The ties between knowledge and information indicated by the quote will be further analyzed in this chapter in order to develop a deeper understanding of the characteristics and relations of these concepts. In spite of a common usage of these terms, their meanings differ significantly between different scientific disciplines (Romhardt 1998). In spite of the many available definitions of knowledge, Weinberger (2012) acknowledges that the consensus within the western culture is limited to very few aspects of knowledge:

"First, knowledge is a subset of belief. We believe many things, but only some of them are knowledge. Second, knowledge consists of beliefs that we have some good reason to believe, whether it's because we've done experiments, because we've proved them logically, [...]. Third, knowledge consists of a body of truths that together express the truth of the world."

By extending this line of thought, Aamodt and Nygård (1995) characterize information and knowledge as *polymorphic* concepts, i.e. concepts that cannot be defined by a classical definition comprehensively describing their specific traits. The definition of a polymorphic concept therefore requires a particular context, which at the same time constitutes the area of validity for the definition. Romhardt (1998) emphasizes the pragmatic approach in various scientific disciplines, where their epistemic[21] interest largely determines the particular semantics for their conception of knowledge.

Although the aforementioned polymorphism of these concepts does not allow universally valid definitions, the study of these concepts within a particular context requires a deeper understanding of the underlying communication processes. Here, *Information Science* (IS) may act as a reference for other scientific disciplines (Beynon-Davies 2009a) in (a) analyzing and providing insights on the process of communication and the involved key elements (Beynon-Davies 2009b), (b) establishing a systematic of conceptual approaches for the definition of the discussed terms (Zins 2007b), and (c) describing the universal characteristics of information systems across time, human cultures, and disciplines (Beynon-Davies 2009b).

---

[21] The adjective *epistemic* refers to knowledge and the degree of its justification (Oxford Dictionaries 2010) in the context of epistemology.

Section 3.1 focuses on the clarification of the concepts and relations of data, information, and knowledge mainly from an information science perspective. The last section presents selected taxonomies for the structuring of knowledge.

3.1 An Information Science Perspective on Data, Information, and Knowledge

As stated in the introduction to this chapter, an information science perspective will be applied to gain sufficient background on the concepts of data, information, and knowledge (D-I-K), their relationships, and underlying communication processes. Due to the high number of available publications on the conceptualization of D-I-K, the present thesis primarily selects review-style articles compiling and structuring the current state of knowledge on this subject. In the first step, a systematic overview of the various approaches for the conceptualization of D-I-K will be established, which will help later on to identify the most appropriate conceptual approach for the present thesis.

Zins (2007b) emphasizes that the concepts of data, information, and knowledge essentially contribute to the theoretical foundations of information science. In the context of a larger endeavor to review and reestablish the theoretical buildings blocks of IS, Zins (2007b) conducted a survey on the definitions and reflections of these concepts among leading IS researchers. The results from this survey formed the basis for a subsequent systematization of the underlying conceptual approaches. Within the resulting taxonomy, the dichotomy of subjective and universal (objective) conceptual approaches constitutes the most significant criteria for the distinction of the definitions on D-I-K:

(a) *Subjective perspective*: From this perspective, D-I-K are perceived as *"inner phenomena bound in the mind of the individual knower"* (Zins 2007b). *Data* are considered as sensory stimuli coupled with an appropriate meaning, whereas *information* is the empirical knowledge[22] gained from the data. *Knowledge* (empirical and non-empirical) is considered as the thought of the individual knower characterized by the person's justifiable belief that it is true[23] (Zins 2007b).

(b) *Universal perspective*: Within the universal domain, D-I-K are considered as *"external phenomena to the mind of the individual knower"* in the shape of human artifacts represented by empirical signs, e.g. printed characters, holes in punch cards, digital signals, or sound waves (Zins 2007b). Hence, in the objective domain data are sets of signs forming an empirical stimulus. Likewise, information is considered as sets of signs representing empirical knowledge, and knowledge is a set of signs representing the content of thoughts (Zins 2007b).

Based on this central distinction, Zins extracted different models for the definition of D-I-K. As it turns out, the most common of these models consists of the definition of data and information as external phenomena and the conceptualization of knowledge as internal phenomena

---

[22] In philosophy, empirical (also: *a posteriori*) knowledge is true belief justified by experience, e.g. perceptual, introspective, and memorial experiences (Steup 2011).

[23] The characterization of knowledge as *"justified true belief"* is one of the classical approaches to define knowledge in philosophy (Steup 2008).

(Zins 2007b). This path of allocating data and information to the universal domain is an approach frequently followed in the literature (Zins 2007a; Beynon-Davies 2009b; Beynon-Davies 2009c; Beynon-Davies 2009a; Probst, Raub, et al. 2010) as it offers the advantage of being valid for both artificial and natural agents in information systems. The conceptualization of knowledge as internal phenomena, however, describes only the mental representations of knowledge. In addition, knowledge bears an objective dimension (VDI 2009; Kebede 2010), which stands for knowledge being represented in an articulated form by means of language or writing.

Concerning the present thesis, obviously both the subjective and objective aspects of knowledge have to be considered when conducting an analysis of knowledge characteristics of MPD. In addition, the intended design support system may be characterized as an information system interacting with both artificial and human agents. Consequently, the applied conceptualization of D-I-K should comprise the concepts for data, information and articulated, explicit knowledge from the universal perspective, whereas tacit knowledge and explicit knowledge in its mental representation are conceptualized within the subjective domain.

Within IS, the knowledge hierarchy, or knowledge pyramid (cf. Figure 3.1), represents a widely accepted model to describe the hierarchical relationships between D-I-K (Nissen 2006; Kebede 2010). The transition between the concepts at the different levels of the knowledge pyramid is commonly perceived as an enrichment process (Romhardt 1998; Kaiser, Conrad, et al. 2008; Conrad 2010; Probst, Raub, et al. 2010). Syntactic rules permit the interpretation of sets of signs from a certain "alphabet" as data that represent objective facts. As data are mainly concerned with the form and representation of symbols for purposes of transmission and storage (Beynon-Davies 2009c), their meaning cannot be interpreted without a particular context of interpretation. Figure 3.1 depicts that the level of *signs* defines the character set used for the representation of the data. At the *data* level, the measured value "117.3 N" stands for the fact represented by the data that cannot be properly understood without an appropriate context of interpretation. Therefore, once data is enriched with context, its content or meaning become interpretable as *information*. The context shown in Figure 3.1 represents the design of a window-lifting system and an associated constraint for the maximal clamping force not to exceed 100 N. The linking and comparison of several pieces of information leads to *knowledge* that directly enables decision-making and taking actions. In the example depicted in Figure 3.1, the design engineer compares the measured value for the clamping force to the constraint and determines that the constraint has been violated for the current design. Consequently, he decides that the current design needs to be altered in order to comply with the restrictions.

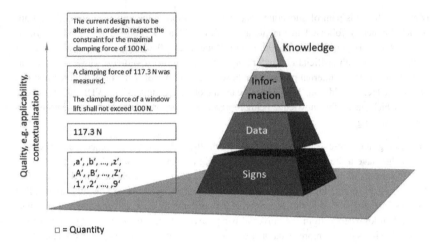

☐ = Quantity

Figure 3.1:   The hierarchy of signs, data, information, and knowledge

In addition, the knowledge pyramid visualizes quantitative and qualitative aspects of the depicted concepts:

(a) The cross-sectional area at each level represents a measure for the *quantity* or *abundance* of the respective concept (Nissen 2006). From a large quantity of signs, a smaller quantity of data is extracted. The data on their part are interpreted to a smaller amount of information, which finally allows gaining even fewer pieces of knowledge.

(b) The height of the pyramid symbolizes *qualitative* aspects of the visualized concepts. When moving upwards from signs to knowledge, the ability to take action, which is called *actionability* or *applicability*, increases (Nissen 2006). Kebede (2010) collected several additional categories falling under the quality dimension, as for instance:

    i. *Degree of distillation*: An increasing degree of distillation is achieved through processes of summarizing, evaluating, comparing, and classifying.

    ii. *Degree of meaningfulness*: Whereas data do not carry any meaning, knowledge is the most meaningful concept in the hierarchy.

    iii. *Level of human input*: When moving higher in the hierarchy, the processes of interpretation, comparison, and linking require a higher level of human input.

So far, the transformation processes between signs, data, information, and knowledge (S-D-I-K) have only been touched briefly. Semiotics, the science of signs and sign systems, provides the necessary theoretical framework to gain a deeper understanding of these processes. Beynon-Davies (2009c) describes four inter-dependent branches of semiotics providing the theoretical means to understand the enrichment process from the layer of signs up to the layer of actions:

(a) *Empirics* studies the physical characteristics of the communication channel transporting a message consisting of signs.

(b) *Syntactics* investigates the principles and rules forming a language from an underlying sign system.

(c) *Semantics* studies the relations between a language's elements and the concepts they refer to. In other words, semantics is the study of the content or meaning carried by a message.

(d) *Pragmatics* focuses on the intention of communication and studies how a recipient achieves extracting the intended meaning of a message that is often ambiguous. The intentions carried by a message establish a link to the layer of actions.

As shown in Figure 3.2, syntactics provides the grammar allowing the extraction of data from signs, whereas semantics establishes the association between data and their incorporated meaning. At the highest level, pragmatics allows extraction of the intentions of the communication and consequently decisions on the actions to take.

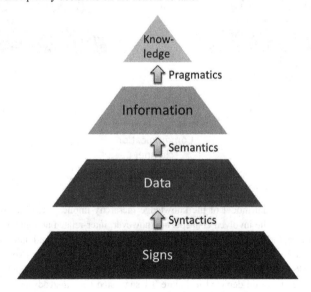

Figure 3.2:   The layers of semiotics in relation to S-D-I-K, after (Bodendorf 2006)

Based on the hierarchical S-D-I-K model, Figure 3.3 depicts the communication process between sender and recipient. First of all, the sender's knowledge is applied to establish a semantic structure representing the information to be transmitted to the recipient (Nissen 2006). In a second step, a message in a particular language encodes the information to be conveyed based on syntax and sign system agreed with the recipient. Next, this message is transmitted as signals through a communication channel, where empirics holds the responsibility for the coding and transmission of the message (Beynon-Davies 2009b). At the recipient's side, the message is transformed through the aforementioned semiotic layers in order to finally enable decision-making and taking actions at the recipient's knowledge level.

Within the described communication process, data and information depend on knowledge that enables the different types of transformation processes (Nissen 2006; Kebede 2010). Nissen (2006) further emphasizes that only signals may be freely flow across time and space as part of the communication process, not data, information, or knowledge. The flows of data, information, and knowledge between people depend on cognitive processes. Information systems may only support the flows of explicit knowledge (Nissen 2006).

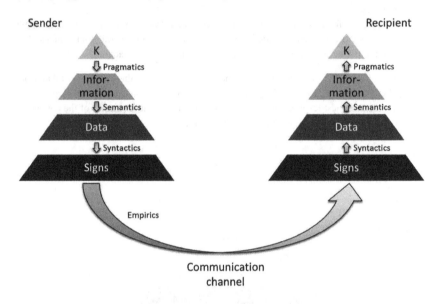

Figure 3.3:   The layers of semiotics within the process of communication

Although the conceptual model of the knowledge hierarchy implies clear hierarchical relations between D-I-K, the involved concepts are, however, inter-related and inextricably interwoven. Therefore, the knowledge hierarchy is likewise perceived as a continuum between the poles of data and knowledge without a strict separation of these concepts (Romhardt 1998; Bodendorf 2006; Zins 2007a; Kebede 2010; Probst, Raub, et al. 2010). The height dimension of the knowledge pyramid depicted in Figure 3.1 suggested such a steady-going change for certain qualitative aspects. Based on the qualitative aspects described by Romhardt (1998) and Kebede (2010), Table 3.1 portraits the value ranges for selected, prominent traits within the continuum from data to knowledge.

Table 3.1:   Prominent traits within the continuum from data to knowledge, based on (Romhardt 1998; Kebede 2010)

Data	Information	Knowledge
simple	◄————————►	complex
concrete	◄————————►	abstract
unstructured	◄————————►	structured
isolated	◄————————►	connected
context-independent	◄————————►	context-dependent
nonmeaningful	◄————————►	meaningful
inapplicable	◄————————►	applicable
low behavioral control	◄————————►	high behavioral control
signs	◄————————►	cognitive behavioral patterns

So far, the essential conceptual approaches and models for the understanding of D-I-K, their relationships, and associated transformation process have been introduced. However, important distinctions on the knowledge level such as the dichotomy of explicit and tacit knowledge were touched on only briefly and will be analyzed more deeply in the following section.

3.2 Knowledge Taxonomies

In order to gain a systematic overview on the different facets of knowledge, the building of knowledge taxonomies from several classification criteria is an appropriate and frequently followed approach. Roth et al. (2010) make clear that due to the polymorphic nature of knowledge, no single, universally valid taxonomy of knowledge may exist. Romhardt (1998) emphasizes the particular value of knowledge taxonomies based on dichotomies, i.e. categories consisting of two opposed terms like *individual* and *collective*. Following this approach, Romhardt extracted 40 different dichotomies proposed for knowledge systematization from different areas of science. The high number of identified categories makes clear that many competing approaches for the systematization of knowledge exist. For this reason, the most relevant categories for knowledge systematization will be identified from the current research literature.

3.2.1 Knowledge Explicitness

Within a philosophical context, Polanyi (1966) introduced the dichotomy of *explicit* and *tacit* knowledge. Nonaka and Takeuchi (1995) extended this model toward the practical application within the context of the organizational knowledge creation (cf. section 5.1.2), where they adopt the continuum between explicit and tacit knowledge as the *epistemological* dimension of knowledge creation. Nissen (2006) introduces the more comprehensible term *explicitness* for the explicit-tacit distinction that will be used in the following.

In the understanding of Nonaka and Takeuchi (1995), tacit knowledge is difficult to articulate and formalize, because it is bound both to an individual and a particular context. In contrast, explicit knowledge refers to knowledge that can be articulated in a formal language and can be transmitted to other people. The different characteristics of these concepts can be best illustrated with the help of a practical scenario described in (Nissen 2006): For the average amateur chef, a recipe written down in a cookbook provides sufficient guidance for the preparation of a meal. The recipe comprises (a) the necessary preparation steps that constitute *explicit* knowledge enabling someone to prepare a meal in combination with (b) the list of ingredients that constitutes required *information* (Nissen 2006). Experienced chefs, however, approach the preparation of a meal in a completely different manner: Over many years, they learned in schools, in practice from head chefs, and by experimentation the principles for blending different ingredients and for their preparation. As a result, expert chefs built up tacit knowledge in cooking enabling them to prepare meals without having to follow written down or memorized recipes. Instead, they observe, smell, and taste the meal under preparation. This way, they determine the appropriate amount of ingredients and the best moment to add them and to finally serve the meal (Nissen 2006).

Table 3.2 contrasts several key characteristics of tacit and explicit knowledge. Tacit knowledge is rooted in experiences and embodied within an individual, whereas explicit knowledge is rational and related to the intellect. Likewise, tacit knowledge is gained within a practical context in an immediate manner, while explicit knowledge is compiled in a sequential mode from individual experiences in the past into an (ideally) context-free theory (Nonaka and Takeuchi 1995).

Table 3.2: The characteristics of tacit and explicit knowledge, according to (Nonaka and Takeuchi 1995)

Tacit knowledge (subjective)	Explicit knowledge (objective)
Knowledge of experience (*body*)	Knowledge of rationality (*mind*)
Simultaneous knowledge (*here and now*)	Sequential knowledge (*there and then*)
Analog knowledge (*practice*)	Digital knowledge (*theory*)

According to Nonaka et al. (2000), the traditional western value system favors explicit over tacit knowledge. In the process of organizational knowledge creation, however, such favoring of one form of knowledge over the other does not reflect the complementary nature of both forms of knowledge (Nonaka, Toyama, et al. 2000; Alavi and Leidner 2001). Section 5.1.2 provides a detailed introduction on the interrelationships of tacit and explicit knowledge within the process of organizational knowledge creation.

In the view of Nonaka and Takeuchi (1995), tacit knowledge possesses cognitive and technical aspects:

(a) *Mental models* (e.g. beliefs, paradigms, perspectives) constitute the *cognitive* aspect of tacit knowledge. Individuals create these mental models as analogies to reality and apply them to improve their understanding of the real world.

(b) The *technical* aspect of tacit knowledge comprises components like know-how, skills, and experiences.

Focussing on the technical aspects of both explicit and tacit knowledge, Snowden (2000) introduces the *ASHEN* (Artifacts, Skills, Heuristics, Experiences, and Natural talent) model that describes where in the continuum between tacit and explicit knowledge the following knowledge categories are located (cf. Figure 3.4):

(a) The technical *artifacts* of knowledge comprise the different pieces of codified knowledge, e.g. documents, databases, procedures, and wikis. Theses artifacts belong clearly to the explicit side of knowledge but may still contain tacit elements of knowledge depending on the success of the codification process.

(b) *Skills* encompass the measurable abilities of how well an individual performs tasks in a particular field. According to Snowden (2000), skills offer the highest potential for codification among the knowledge resources of an organization.

(c) The term *heuristics* describes experience-based methods (i.e. so called "rules of thumb") that rapidly produce results in situations where time and/or complexity constraints prevent the application of more generally valid and precise methods. Snowden (2000) emphasizes the potential of identifying and codifying heuristics to make this valuable, experience-based knowledge available to others.

(d) Snowden (2000) perceives *experiences* as the most valuable and most difficult of the tacit assets of an organization. The difficulties result from the facts that (a) experiences arise from working under particular situations, and (b) the replication of these situations may be difficult. These issues impede the direct sharing and transfer of experiences.

(e) *Natural talent* is the most tacit knowledge category. It is bound to an individual and cannot be transferred to others.

In addition to the widely accepted concepts of explicit and tacit knowledge, some authors (Nickols 2000; Ahmed, Hacker, et al. 2005; Nissen 2006) introduce *implicit* knowledge as a third knowledge type within the *epistemological* dimension. Unfortunately, the conceptualization of implicit knowledge differs significantly between the authors:

- Ahmed et al. (2005) contrasts tacit knowledge, which cannot be articulated by the individual possessing it, by implicit knowledge that is derived and articulated from tacit knowledge by different people. She locates implicit knowledge between tacit and explicit knowledge.

- Accordingly, Nickols (2000) places implicit knowledge between tacit and explicit knowledge. In contradiction to Ahmed, however, he perceives implicit knowledge as knowledge that can be articulated but has not been, and contrasts it with tacit knowledge that cannot be articulated at all.

- Nissen (2006) on his turn, views implicit knowledge in the same way as Nickols as articulable knowledge that has not been articulated yet. In contrast to Ahmed and Nickols, though, he perceives implicit knowledge as a subset of tacit knowledge.

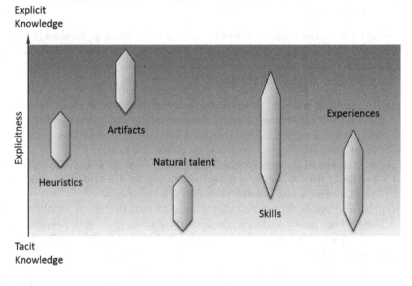

Figure 3.4:   The ASHEN model of knowledge categories, according to VDI guideline 5610 (2009)

Apart from the many differences described in the conceptualization of implicit knowledge, the authors agree in the positioning of implicit knowledge between tacit and explicit knowledge. A widely accepted definition for implicit knowledge, however, remains to be established[24]. Regarding the conceptualization of tacit and explicit knowledge, the generally accepted definitions introduced by Nonaka and Takeuchi (1995) will be adopted within the present thesis, which are outlined in the following:

Definition 3.1:   Tacit knowledge, according to (Nonaka and Takeuchi 1995)

> Tacit knowledge refers to a form of knowledge that is difficult to articulate and formalize through means like language or writing, because it is bound both to an individual and to a particular context.

---

[24] In the context of cognitive psychology, though, implicit knowledge commonly represents unconscious knowledge equivalent in many ways to tacit knowledge. Dienes and Perner (1999) for instance propose a representational theory of implicit and explicit knowledge. Within this theory, they distinguish implicit and explicit knowledge based on the representation of the content, the attitude adapted for the justification of knowledge, and the role of the self within justification.

Definition 3.2: Explicit knowledge

Explicit knowledge refers to a form of knowledge that can be articulated through words, diagrams, formulae, computer programs, and similar means and can be readily transmitted to other people. Explicit knowledge can either be represented in the form of mental representations in the human brain or in its physical form by means of language or writing.

3.2.2 Organizational Reach of Knowledge

The briefly mentioned model for organizational knowledge creation introduced by Nonaka and Takeuchi (1995), encompasses the different organizational levels of knowledge creation as a second, the so called *ontological* dimension. The different levels of knowledge creation start at the individual level, where individuals create and embody personal knowledge. This personal knowledge is then amplified at increasing levels of organizational interaction, reaching from groups, to organizations and finally crosses inter-organizational level boundaries (cf. section 5.1.2).

Alavi and Leidner (2001) likewise mention the distinction of individual and social knowledge within their comprehensive overview on knowledge taxonomies. Nissen (2006) introduced the more comprehensible term *reach* to describe the level of social aggregation within the model for organizational knowledge creation. This term will be adopted in the following.

3.2.3 Causal, Conditional and Strategic Knowledge

In addition to the already introduced knowledge categories, Alavi and Leidner (2001) mention supplementary classification criteria, like *causal* knowledge (*"knowing why"*) and *conditional* knowledge (*"knowing when"*). Ahmed, Hacker, et al. (2005), however, favor the designation of *situational* knowledge for conditional knowledge. Drawing from Venselaar, van der Hoop, et al. (1987), Ahmed et al. introduce the notion of *strategic* knowledge that comprises the various aspects of knowledge on processes (e.g. algorithms, heuristics) relevant in problem solving.

3.2.4 Declarative and Procedural Knowledge

A widely applied dichotomy originating from cognitive psychology contrasts declarative and procedural knowledge (Nickols 2000; Conrad 2010). *Declarative* knowledge is characterized as *"knowing that"* and encompasses the knowledge of facts, concepts, methods, and procedures (Nickols 2000; Nissen 2006). In contrast, *procedural* knowledge focuses on *"knowing how"* to apply the different aspects of declarative knowledge in an applicative context (Nickols 2000; Conrad 2010).

Nonaka and Takeuchi (1995) consider this knowledge dichotomy synonymous to their distinction of tacit and explicit knowledge. In their interpretation, *declarative* knowledge corresponds to *explicit* knowledge, whereas *declarative* knowledge matches to *explicit* knowledge. Due to the adoption of the model of organizational knowledge creation (cf. section 5.1) by the present thesis, it seems advisable to adhere to its vocabulary of concepts. In the following, the concept of *explicit* knowledge will therefore be considered synonymous to *declarative* knowledge as well as *tacit* knowledge synonymous to *procedural* knowledge.

3.2.5 Conclusions

So far, the most relevant categories for knowledge systematization have been introduced based on the current research literature. Figure 3.5 depicts these classification criteria for knowledge taxonomies in combination with possible values for each criterion.

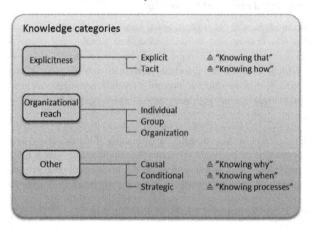

Figure 3.5: Hierarchy of criteria forming different knowledge taxonomies

Alavi and Leidner (2001) emphasize that in many cases a pragmatic approach for the classification of knowledge valuable for a particular organization will be adopted. Following such a pragmatic approach, the various types of knowledge encountered within the business processes of an organization will be taken into account and may encompass e.g. knowledge of customers, competitors, products, and processes (Alavi and Leidner 2001). This line of thought will be extended in section 6.2 toward specific knowledge taxonomies within product development.

4 A Cognitive Psychology Perspective on Individual Knowledge Creation

> *If Freud is right, understanding, inference, conscious choice, and the like play much less of a role in human behavior than his more rationalistic contemporaries thought. Although I am hardly a fan of Freud, I have a fairly skeptical attitude toward the role of these factors in human affairs. The rule that human beings seem to follow is to engage the brain only when all else fails – and usually not even then.*
>
> David L. Hull, 2001

In the above quote, Hull (2001) humorously illustrates the concept of humans as *cognitive misers* (Stanovich 2009). According to this concept, human beings conduct typically the least possible level of cognitive activities to cope with a given situation in order to preserve the limited cognitive resources. Consequently, they rely on intuitive decision-making as long as the given context permits and apply reflective reasoning only if required. Dual-process theories originating from cognitive psychology provide the descriptive models that are able to explain such behavior. In the context of product development, dual-process theories seem promising to explain the mutual relationships of intuition and logical reasoning in decision processes and to understand the types of knowledge involved in the cognitive activities of individuals when developing products. In particular, two types of cognitive processes are of major interest for knowledge creation in individuals: First, an individual may arrive at new insights typically by means of *reasoning* - applying for instance deductive inferences - or by *creative thinking* - involving abductive inferences and strategies for problem solving. Secondly, the insights gained may become part of the long-term memory of an individual by means of *learning*. In addition, an individual may also learn implicitly (i.e. unconsciously) while interacting with its environment during experiments. Moreover, the individual can learn already existing knowledge from other individuals during an apprenticeship or training, or explicitly from external sources of knowledge (e.g. books, scientific papers, blog entries, etc.).

Overall, *cognitive psychology* provides the theories explaining the processes of lower-level cognition (perception, attention, and memory) and higher-level cognition (thinking and reasoning), focusing strongly on empirical evidence. Its conceptualizations and research results have been an influential source for the scientific areas of e.g. organizational knowledge creation and design cognition. Although Nonaka and Takeuchi (1995) mention the influences of cognitive psychology on the formation of their theory only in passing, in a later publication Nonaka and von Krogh (2009) admit that modern cognitive psychology provides theoretical and empirical insights into the interaction of tacit and explicit knowledge in individuals.

Overall, thinking is the most general term for mental activities such as conceptualizing, reasoning, deciding and planning (Holyoak and Morrison 2012). Holyoak and Morrison define thinking as *"the systematic transformation of mental representations of knowledge to characterize actual or possible states of the world, often in service of goals."*

Section 4.1 gives an overview on the various approaches to thinking and reasoning and explains the relationships of the descriptive models to normative theories. Next, section 4.2 provides an introduction on dual-process theories of thinking processes. Finally, section 4.3 introduces the tripartite model of mind that will be applied throughout the present thesis. This model allows explanation of the connection of two types of cognitive processes to tacit and explicit knowledge and the associated learning processes.

4.1 Overview on Descriptive Models of Thinking

At present, cognitive psychology does not provide a unique, coherent body of theory that explains the relevant traits of higher-level cognition. Instead, the different phenomena of thinking are studied in several interrelated scientific fields (Goel 2007; Holyoak and Morrison 2012):

(a) *Reasoning* focuses on the cognitive processes of drawing a conclusion from given information and background knowledge.

(b) *Decision-making and judgment* study the cognitive processes of choosing one option over possible alternatives.

(c) *Problem solving* analyzes the cognitive processes and in particular the actions conducted to achieve a goal, where the *problem* is defined as the mental representation of the overall task to accomplish the goal (Visser 2006a).

This fragmentation of perspectives on thinking leads to a large overlapping between these scientific fields and hinders the use of these theories for real-world problems that typically involve more than one of the above phenomena.

The research of the present thesis is directed toward understanding the cognitive processes utilized for individual knowledge creation in the context of mechatronic product development at a rather coarse level. The modeling of the cognitive activities should only capture the characteristics of (a) the information and knowledge artifacts used as input and transformed in this cognitive activity, (b) the utilized background knowledge, (c) the generated information and knowledge objects at the output, and (d) a coarse description of the conducted cognitive activity distinguishing it from others (cf. Figure 4.1).

An approach frequently encountered in descriptive theories of thinking consists in the adoption of normative theories (logic, probability, rational choice) as reference point for the theoretical frameworks (Chater and Oaksford 2012). For instance, logical inferencing is commonly applied in the context of reasoning but perceived at least as controversial and by proponents of some research approaches as a fundamental mistake (Chater and Oaksford 2012). Theoretical frameworks in decision-making and judgment often apply probability theory and rational choice.

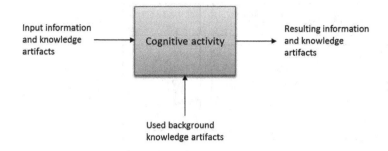

Figure 4.1: Model for cognitive activity in the context of individual knowledge creation

Section 6.1.1 assesses for the context of product development, which of the above-mentioned phenomena and associated approaches are best suited to the description of the cognitive activities in product development.

4.2 Dual-Process Theories of Thinking

Dual-system theories of cognitive psychology have been developed over the last 30 years (Evans 2008). During the last 10 years, however, they have gained more attention partly due to the Nobel Prize in economics awarded to the psychologist Daniel Kahneman for the prospect theory leading to behavioral economics. Kahneman and Tversky developed a dual-process theory for judgment and decision-making that constitutes the basis for the above mentioned prospect theory (Kahneman 2003). At the present, however, there is no single, consolidated dual-process theory but various dual-process theories proposed by different researchers in reasoning, and decision-making and judgment. In the following, the two-system theory of Kahneman and Tversky for decision-making and judgment will be presented merely as an introduction into the structure of a typical dual-system theory. Later, the reflections of Evans and Stanovich (2013) on criticism toward dual-process theories will be employed to generalize the characteristics of dual-processing theories confirmed by empirical evidence.

4.2.1 Dual-System View in Decision Making and Judgment

Kahneman and Tversky developed their dual-system view from experiments on intuitive judgment of experts: They encountered systematic errors between the statistical intuition and the statistical knowledge of researchers although all of them were well experienced in statistics. From there, they developed a dual-system view consisting of separated systems for intuition and for reasoning responsible for the two modes of thought (Kahneman 2003; Kahneman 2011). Stanovich and West (2000) coined the terms *System 1* and *System 2* for these cognitive systems.

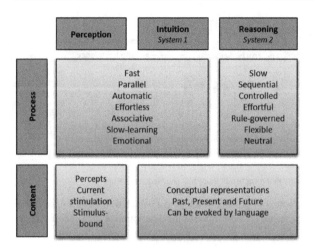

Figure 4.2:   Process and content of two cognitive systems (Kahneman 2003)

Figure 4.2 illustrates the characteristics of the two types of cognitive processes in the two-system view of Kahneman (2003) as well as their content and the responsible cognitive systems (Perception, System 1, and System 2). The processes conducted by System 1 and perception share many commonalities as both run in a fast, automatic, effortless, and rather subconscious manner. In contrast, the processes of System 2 require effort, run slowly and sequentially, and can be consciously controlled. In summary, System 1 processes can be labeled as intuitive and emotional, whereas System 2 processes are considered as deliberative and logical. In spite of the similar process characteristics of perception and System 1, the content treated by these processed largely differs: Whereas perception handles only the current stimulation, both System 1 and 2 focus on conceptual representations of the past, present or future that can be evoked by language.

Kahneman (2003) distinguishes processes adhering to System 1 and 2 based on their concurrency: Whereas the intuitive processes of System 1 do not require explicit attention and run without noticeable effort, the reasoning processes of System 2 require an explicit mental effort that requires rare working memory. Therefore, the effortful processes of System 2 tend to disrupt each other and run in serial mode (Kahneman 2003).

The intuitive thoughts developed by System 1 are usually subject of permanent control and correction by System 2. If an individual is overcharged with effortful cognitive activities, however, System 2 will cease to efficiently control the judgments developed by System 1. In such situations, it may occur that intuitive thoughts and judgments are expressed without the usual validation and correction of System 2. Altogether Kahneman (2003) distinguishes five ways in which a decision can be taken:

(1) System 1 provides an intuitive judgment that is
    (a) Confirmed by System 2, or
    (b) Extended by other relevant aspects, or

(c) Recognized by System 2 as containing a bias and accordingly corrected, or

(d) Contradicting a respected rule and suppressed.

(2) System 1 does not possess sufficient knowledge to provide an intuitive response and therefore System 2 has to reason on a valid response.

The dual-system model of Kahneman and Tversky also enables an explanation to the commonly observed phenomena that experts with many years of experience in a certain field arrive at intuitive judgments that surprise even other people knowledgeable in the particular field. Kahneman (2011) provides some impressive illustrations of such *expert intuition*: He describes the story of a firefighter commander who suddenly urges his colleagues to leave the scene of fire just before the wooden floor collapsed. Although the firefighter did not have sufficient evidence of an invisible fire underneath, he intuitively grasped that something was wrong. Kahneman (2011) explains expert intuition by intuitively recognizing familiar elements in a new situation, an activity clearly associated with System 1. In contrast, he explains intuitive judgments by some *heuristics* applied by System 2.

In summary, Kahneman (2011) describes the two systems as collaborating agents with specific functions, strengths, and weaknesses. The combination of System 1 and System 2 leads to a highly efficient cognitive system that minimizes effort and optimizes performance (Kahneman 2011).

4.2.2 Generalization of Dual-Processing and Dual System Theories

Evans and Stanovich (2013) observed that the gains in popularity for dual-process and dual-system theories in recent years have been accompanied by an increased level of criticism toward these theories. Therefore, they reviewed the attributes proposed by various dual-processing and dual-system theories and extracted and consolidated contributions based on the available empirical evidence. As the result of their systematization, they remain skeptical toward the proposed association of the two types of cognitive processes (intuitive vs. reflective) with exactly two cognitive systems (System 1 and System 2) as they state sufficient evidence for the association of each processing type with multiple cognitive systems. Therefore, they propose using the notion of Type 1 and Type 2 processing instead, as this allows multiple cognitive systems for each type of processing.

The top part of Table 4.1 shows the characteristics commonly attributed to the Type 1 and Type 2 processes from dual-process theories. The lower section of the table depicts additional attributes associated with the two systems originating from dual-system theories, where System 1 represents the evolutionary old mind common to both animals and humans, whereas System 2 corresponds to the uniquely human mind.

Evans and Stanovich (2013) perceive *autonomy* as the unique feature of Type 1 processing. They associate processes of associative and implicit learning as well as conditioning with this processing type. In contrast, they define *cognitive decoupling* as a key feature defining Type 2 processing. Cognitive decoupling is perceived as the ability to generate a secondary representation from the primary representation of the given context that is decoupled from the world (Stanovich, West, et al. 2011). The secondary representation can be employed for such Type 2 processes as mental simulations and hypothetical reasoning. It is obvious that these processes are a key element required for creative thinking in product development that equally depends on the previously described expert intuition.

Table 4.1: Attributes of dual-process and dual-system theories of higher cognition, according to (Evans and Stanovich 2013)

Type 1 process (intuitive)	Type 2 process (reflective)
Defining Features	
Does not require working memory	Requires working memory
Autonomous	Cognitive decoupling: mental simulations
Typical correlates	
Fast	Slow
High capacity	Capacity limited
Parallel	Serial
Nonconscious	Conscious
Biased responses	Normative responses
Contextualized	Abstract
Automatic	Controlled
Associative	Rule-based
Experience-based decision making	Consequential decision making
Independent of cognitive ability	Correlated with cognitive ability
System 1 (old mind)	**System 2 (new mind)**
Evolved early	Evolved late
Similar to animal cognition	Distinctively human
Implicit knowledge[25]	Explicit knowledge
Basic emotions	Complex emotions

4.3 Tripartite Model of Mind

The generalization of dual-process theories presented in the previous section presupposes that each of the two processing types can be conducted by multiple cognitive systems. The perspective presented here on dual-processing theories, however, does not clarify the nature of the involved cognitive systems, their mutual relationships, or the allocation of cognitive systems to each processing type. Here, the tripartite model of mind introduced by Stanovich, West, et al. (2011) represents an excellent starting point as (a) it takes into account the generalized ideas of dual-process theories, (b) identifies several cognitive systems, (c) allocates them to each type of cognitive processing, (d) describes the relationships between the cognitive systems and (e) identifies the knowledge structures of each cognitive system.

---

[25] In the context of cognitive psychology, implicit knowledge commonly represents unconscious knowledge equivalent in many ways to tacit knowledge.

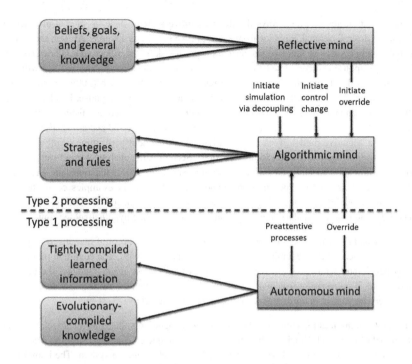

Figure 4.3: The tripartite model of mind associated with two types of processing, adapted from (Stanovich 2009; Stanovich, West, et al. 2011)

Figure 4.3 gives an overview on the relevant constituents of the tripartite model of mind: In the lower part of Figure 4.3, the autonomous mind represents a heterogeneous set of cognitive systems conducting Type 1 processes, for which Stanovich (2009) proposed the acronym TASS (The Autonomous Set of Systems). TASS contains evolutionary-compiled knowledge encapsulated in knowledge bases as well as knowledge learned to automaticity (e.g. rules, procedures, decision-making principles). Both types of knowledge applied in TASS can be characterized as *tacit* due to their subconscious character. TASS uses processes of associative and implicit learning for acquiring knowledge (Stanovich 2009).

In the upper part of Figure 4.3, the algorithmic and reflective minds represent the cognitive systems conducting Type 2 processing. Stanovich (2009) introduced the distinction of the algorithmic and the reflective minds to distinguish the *cognitive ability* usually measured in tests of intelligence from the *ability for critical or rational thinking*. The algorithmic mind is positioned as subordinate to the reflective mind that regulates the algorithmic mind according to individual and epistemic goals. For this purpose, the reflective mind utilizes general knowledge and strategies as well as the individual's beliefs and goals (Stanovich, West, et al. 2011). In contrast, the algorithmic mind employs micro-strategies for steering the cognitive activities and rules for the sequencing of behaviors and thoughts (Stanovich, West, et al. 2011).

Figure 4.3 indicates in which ways the three cognitive systems have to interact in order to obtain the desired rational behavior from the cognitive system as a whole: At first, the autonomous mind provides an intuitive response for a given task. If the reflective mind detects that the given response is incomplete or violating any of the personal beliefs and goals, it will initiate an override of the intuitively obtained response. Here, Stanovich (2009) proposes two ways in which the analytical system can produce a more satisfying response. Firstly, the analytical mind can employ *serial associative cognition* – a cognitive activity distinct from hypothetical thinking. It stands for a mode of thinking conducted in a serial and associative manner based on a cognitive model given to the individual and without performing a full cognitive decoupling from the real world (Stanovich 2009). Secondly, the analytical mind can perform a mental simulation based on cognitive decoupling. Stanovich (2011) illustrates the difference between serial associative cognition and hypothetical thinking in examples contrasting an associative mode of thinking, where one thought leads to the next, with strategic planning activities requiring a secondary representation maintained during hypothetical reasoning.

The described tripartite model of minds conforms well to the concept of humans as *cognitive misers* introduced at the beginning of the chapter (Stanovich 2009). The cognitive processing starts per default with the cognitive activity implying the least effort: First, try the autonomous mind, if no response was given or an override is required, try serial associative reasoning, otherwise engage in a full mental simulation (Stanovich 2009).

Table 4.2 summarizes the characteristics of the three cognitive systems in Stanovich's tripartite model of mind and associates the attributes of the conducted cognitive activities, the types of learning, and the employed types of knowledge to each cognitive system. The knowledge employed by the analytical mind can be characterized as explicit due to its rather conscious character and due to the possibility to articulate them.

Overall, Stanovich's tripartite model of mind introduces a novel approach to understand the cognitive activities conducted in reasoning, decision-making and judgment based on empirical and theoretical evidence. It offers a fascinating perspective to gain a deeper understanding of:

(a) The cognitive activities conducted in design and more general in product development,
(b) The types of knowledge typically applied in these cognitive operations,
(c) The dichotomy of rational thinking and expert intuition in design,
(d) The specifics of the learning activities for each type of cognitive activity, and
(e) The importance of hypothetical reasoning and cognitive simulations for creativity in design.

Table 4.2:   Characteristics of the three types of cognitive systems in Stanovich's tripartite model of mind

Cognitive system	Characteristics of cognitive activities	Types of learning	Types of knowledge	
Autonomous mind (TASS)	Preattentive	Associative and implicit learning	Tacit	Knowledge learned to automaticity
	Heuristic			Evolutionary-compiled knowledge
	Intuitive			
Algorithmic mind	Serial associative cognition	Explicit and implicit learning	Explicit	Micro-strategies for cognitive activities and rules for sequencing
	Hypothetical reasoning			
Reflective mind	Regulation based on personal beliefs and epistemic goals			Beliefs, goals and general knowledge and strategies

5 Organizational Knowledge Creation

The organisation is not merely an information processing machine, but an entity that creates knowledge through action and interaction. It interacts with its environment, and reshapes the environment and even itself through the process of knowledge creation.

Nonaka, Toyama et al., 2000

In the above quote, Nonaka, Toyama, et al. (2000) lay out their (then) unorthodox understanding of an organization as an institution centered on the creation and amplification of knowledge. This view strongly contrasts to the perception of traditional Western management of an *"organization as a machine for 'information processing'"* (Nonaka and Takeuchi 1995). From these contradictory views on this topic in the Western and Japanese culture, Nonaka and Takeuchi (1995) depart to develop a theory of organizational knowledge centered on the social interplay of tacit and explicit knowledge. They describe four sequential knowledge conversion modes in a two-dimensional system formed by (a) tacit and explicit knowledge and by (b) the various organizational levels (individual, group, organization). Even its critics (Gourlay 2006) admit the quasi-paradigmatic status of the this theory and recognize it as one of the most influential approaches in organizational learning and knowledge management.

Section 5.1 provides an introduction on this theory comprising knowledge flows, knowledge contexts, and knowledge resources. Subsequently, section 5.2 introduces the aims, components, and strategies of knowledge management (KM) followed by critical reflections on the current conception and practice of KM.

5.1 A Knowledge Science Perspective on the Model of Organizational Knowledge Creation

In the previous chapter, some groundwork has been laid for the understanding of the components (D-I-K) and for the classes of knowledge involved in business processes like planning, decision-making, and problem solving. On this basis, the processes of organizational knowledge creation and amplification in combination with the knowledge flows will be introduced in the following. In terms of the ternary model of organizational knowledge creation, these concepts are subsumed under the term (a) *knowledge dynamics* (Kusunoki, Nonaka, et al. 1998; Nonaka, Toyama, et al. 2000). The model's complimentary two elements consisting of (b) *knowledge base* or *knowledge resources*, and (c) *knowledge frame, knowledge context* or as expressed in Japanese *"ba"*[26] will be covered subsequently.

Today, some researchers (Nissen 2006; Zins 2007a; School of Knowledge Science 2010; Nakamori 2011) believe that there is a need for a new discipline called *knowledge science* (KS). Nissen (2006) emphasizes the need for such an interdisciplinary field drawing from information science, social sciences, and physical sciences due to the *"confusion of current KM research"*. In Japan, though, the *School of Knowledge Science* at the Japan Advanced Institute of Science and Technology (JAIST) has brought together scientists from the many disciplines involved in the interdisciplinary research on knowledge creation processes, decision making processes, and the overall nature of knowledge since 1998 (School of Knowledge Science 2010; Nakamori 2011). This school was started based on the theory on organizational knowledge creation of Ikujiro Nonaka, who also served as the first dean of the school (Nakamori 2011).

Altogether, knowledge science is perceived as a problem-oriented interdisciplinary field that focuses on the modeling of knowledge formation and communication processes in organizations (School of Knowledge Science 2010; Nakamori 2011). Drawing from the principles and research results from knowledge science, *knowledge technology* provides technological support within knowledge creation, knowledge discovery and knowledge management (Nakamori 2011).

Figure 5.1 gives an overview of four research areas sharing common research topics with knowledge science and their approaches. Within the four approaches, italic letters indicate steps applying tacit knowledge. *Knowledge engineering* aims at the practical application of methods from a*rtificial intelligence* through the elicitation of an expert's knowledge and its subsequent application within a knowledge-based system. In the first step, *knowledge discovery* aims at the extraction of partial rules (relationships and patterns) from a given data set. Based on domain knowledge, the meaning of the discovered patterns will be extracted resulting in new knowledge. *Knowledge construction* is an approach for knowledge creation starting with a simulation of complex phenomena based on an initial hypothesis and the subsequent adding of meaning to emerged properties (Nakamori 2011). *Knowledge management* comprises the purposeful and methodical management of knowledge creation, storage, sharing, and its application (Kebede 2010).

---

[26] The Japanese word *„ba"* stands roughly for the word „place" and describes the shared context in which knowledge is created, shared, and applied (Nonaka, Toyama, et al. 2000). Nonaka et al. (2000) perceive this term as more than just a physical location: They conceptualize it as (a) a point in space and time, a so called "time-space nexus", and (b) as the super ordinate concept for physical, virtual, and mental spaces. Moreover, they consider *ba* as the context where the linking and comparison of several pieces of information leading to knowledge takes place.

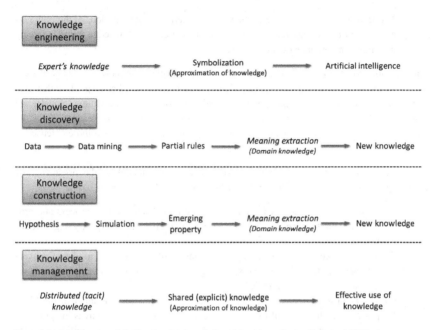

Figure 5.1: Existing research fields related to knowledge science, according to (Nakamori 2011)

The following three sections draw from research results on the model of organizational knowledge creation developed mainly by researchers from the aforementioned *School of Knowledge Science* at JAIST.

5.1.1 Organizational Knowledge Creation: Process and Model

Nonaka and Takeuchi (1995) developed a model for organizational knowledge creation from their finding of strong differences between the Western and Japanese culture on knowledge creation and the associated value system for explicit and tacit knowledge. Within the traditional Western management, there prevails the conception of an *"organization as a machine for 'information processing'"* (Nonaka and Takeuchi 1995). This position leads to a preference for explicit knowledge and does not permit an explanation of how knowledge is created. Japanese companies, however, perceive knowledge primarily as tacit. From these opposing views, Nonaka and Takeuchi (1995) develop their understanding of organizational knowledge creation as a two-dimensional interplay of tacit and explicit knowledge (*explicitness* dimension) at different organizational levels (*organizational reach* dimension).

From this initial model that only covers the aspects of *knowledge dynamics*, Kusunoki, Nonaka et al. (1998) developed a larger conceptual framework that comprises, in addition to knowledge dynamics, two other types of organizational capabilities:

(a) The *knowledge base* comprises all units of knowledge present at the different levels of an organization.

(b) The *knowledge frame* represents the various patterns or configurations of the individual pieces of knowledge forming the knowledge base. The knowledge frame is largely determined by organizational structures.

Subsequently, Nonaka and Konno (1998) refined this conceptual framework through the extension of the *knowledge frame* by introduction of the aforementioned concept of *ba* that may be perceived as a frame for the process of knowledge creation in space and time (Nonaka and Konno 1998). Later on, Nonaka et al. (2000) included questions of the leadership for the knowledge-creating process in their model, an aspect that has been studied more extensively by von Krogh, Nonaka et al. (2012).

Figure 5.2 depicts the three constituents of the conceptual model of the organizational knowledge-creating process and their relationships (Nonaka, Toyama, et al. 2000):

(a) *Ba* acts as the creative environment for the knowledge creation process and provides it with the necessary space, energy (e.g. motivation) and quality (e.g. level of standards).
(b) The actual knowledge creation process consists of four different modes of conversion between tacit and explicit knowledge: *socialization, externalization, combination,* and *internalization* (SECI).
(c) The *knowledge resources* supply the necessary knowledge as input to the SECI process. As the result of the SECI process, the knowledge resources integrate the newly created knowledge. In addition, they moderate how *ba* may fulfill its role as platform for the knowledge creation process.

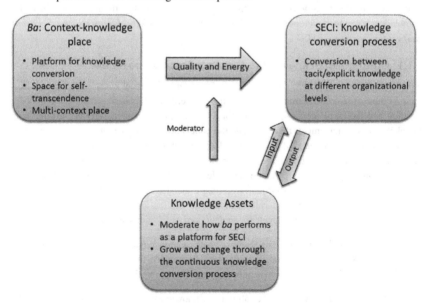

Figure 5.2: Conceptual model of knowledge-creating process, according to (Nonaka, Toyama, et al. 2000)

In the next sections, the three constituents of the process of organizational knowledge creation will be presented in more details based on (Nonaka and Takeuchi 1995; Nonaka and Konno 1998; Nonaka, Toyama, et al. 2000; Nissen 2006; von Krogh, Nonaka, et al. 2012).

5.1.2 Knowledge Dynamics: Knowledge Flows and Life Cycles

With the SECI process (cf. Figure 5.3), Nonaka and Takeuchi (1995) provide a conceptual model for the process of organizational knowledge creation that is based on the interplay of tacit and explicit knowledge within the four distinct knowledge conversion activities. As the result of these conversion activities, the organizational knowledge resources are (a) extended by the newly created knowledge and (b) made accessible to a larger group of people within an organization (Nonaka and Takeuchi 1995).

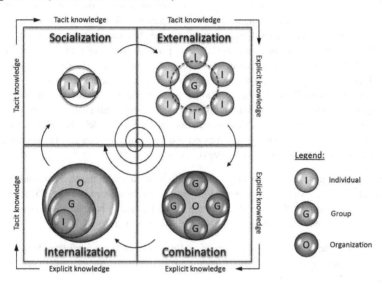

Figure 5.3: The SECI process of organizational knowledge creation, according to (Nonaka and Konno 1998)

In the following, the four modes of knowledge conversion will be briefly introduced (Nonaka and Konno 1998; Nonaka, Toyama, et al. 2000):

(a) *Socialization* covers the sharing of tacit knowledge between individuals. Tacit knowledge may be exchanged only through shared experiences resulting from joint activities in combination with physical proximity, e.g. during an apprenticeship or pair programming. Knowledge sharing may also occur outside the working environment, where complementary facets of tacit knowledge (e.g. mental models, mutual esteem, and trust) are created and communicated during informal gatherings.

(b) *Externalization* describes the conversion of tacit knowledge into explicit concepts, which are comprehensible to a larger group. Nonaka and Takeuchi (1995) emphasize the importance of the reasoning methods *deduction*, *induction*, and *abduction* for this conversion mode as they support establishing shared explicit concepts and consequently generate new knowledge. In cases where these analytical methods do not provide sufficient support for the formulation of explicit concepts, dedicated techniques for the transformation of tacit knowledge into explicit forms using metaphors, analogies, schemas, and dialoguing are required (Nonaka and Takeuchi 1995).

(c) *Combination* refers to the process of converting explicit knowledge into new and more complex sets of explicit knowledge. Nonaka and Konno (1998) emphasize that combination relies on three distinct sub-processes:

    i.   Capturing and compiling explicit knowledge from internal and external sources.

    ii.   The communication of these newly created sets of explicit knowledge to other groups within the organization (cf. Figure 5.3).

    iii.   The editing, sorting, classifying and combining of various artifacts of explicit knowledge makes them more easily understandable for groups from other contexts, and may itself create new knowledge (Nonaka and Takeuchi 1995).

(d) *Internalization* describes the embodiment of parts of a company's explicit knowledge within the shared tacit knowledge resources of an organization at various levels (cf. Figure 5.3). Evidently, each individual has to internalize the relevant pieces of codified knowledge by him or herself. On the organizational level, however, trainee programs or activities such as learning-by-doing may support this process and assure that the individuals share the relevant perceptions and mental models. In addition, virtual environments permitting experimentation and simulation may support the internalization process (Nonaka and Konno 1998).

The process of organizational knowledge creation does not stop with the *internalization* step. As indicated by the arrow between *internalization* and *socialization* in Figure 5.3, the tacit knowledge acquired from the internalization step may subsequently be shared with others and lead to a new turn of the knowledge spiral (Nonaka, Toyama, et al. 2000).

Nissen (2006) extends Nonaka's two-dimensional model of organizational knowledge creation by two complementary dimensions leading to the four-dimensional model depicted in Figure 5.4:

(a) The *life-cycle* dimension describes the *knowledge management* (KM) activities taking place during a particular knowledge flow process. As indicated in Figure 5.4, Nissen selects for his model six KM activities of the many proposed core activities by KM frameworks. Section 5.2.1 provides more details on core activities in various KM frameworks.

(b) The *flow time* captures the time required for the flow of knowledge between the considered participants at their organizational levels. Here, the thickness of the arrow symbolizes the flow time: Thin arrows visualize a short flow time, whereas thick arrows stand for long-lasting knowledge flows (cf. Figure 5.4).

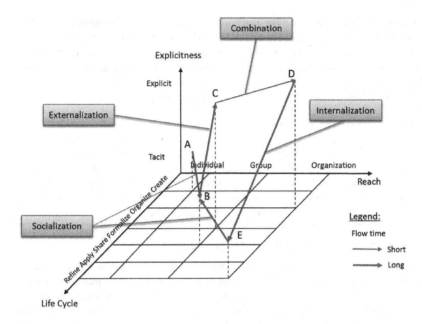

Figure 5.4: Four-dimensional model of knowledge flows, according to (Nissen 2006)

The diagram in Figure 5.4 depicts Nonaka's SECI process complemented the two supplementary dimensions. Within the first *socialization* activity (A→B), tacit knowledge is shared and amplified from the individual to the group level. Within the life cycle dimension, this activity is characterized as *"share"*. The subsequent *externalization* step (B→C) transforms and formalizes tacit knowledge that reemerges as explicit knowledge remaining within the same group. The KM activity *"formalize"* aptly describes this externalization step. Within the following *combination* activity (C→D), the explicit knowledge amplifies its organizational reach from the group to the organization level. In contrast to the preceding steps, the flow time is shorter as it involves explicit knowledge only. At the life cycle scale, the *"organize"* activity captures best the structuring character of the involved activities. The subsequent *internalization* step (D→E) describes the embodiment of parts of the company's explicit knowledge within the shared knowledge resources of an organization. Therefore, the arrow from D to E describes the conversion from explicit into tacit knowledge remaining on the same organizational level. Within the life cycle dimension, this activity is characterized as *"refine"*. Finally, Nissen concludes the cycle with a second *socialization* activity (E→B) covering the learning of the shared tacit knowledge resources of an organization by groups and individuals.

So far, the term *"knowledge flow"* has been used without formally introducing it. According to Nissen (2006), this term refers to the movement of knowledge between coordinates, e.g. organizational reach, or points in space and time. Zhuge (2004) adopts a rather generalist definition by characterizing a knowledge flow as the passing of knowledge between people and machinery. Holsapple and Joshi (2004) perceive a knowledge flow as the transfer of knowledge between two instances of KM activities along the way the associated knowledge representation may change, a view that focuses on the life cycle dimension of the four-dimensional model depicted in Figure 5.4. Similarly, Newman (2003) defines a knowledge flow as a sequence of transformations of knowledge artifacts by the various agents involved in the process. The four-dimensional model of knowledge flows (Nissen 2006) covers all of these definitions and will therefore be applied as the theoretical basis for establishing a comprehensive definition drawing from the aforementioned conceptualizations:

Definition 5.1:   Knowledge flow, drawing from (Nissen 2006)

A knowledge flow refers to the transfer of knowledge between different participants as sender and receiver each affiliated with an organizational level (individual, group, organization). A knowledge flow may transform the representation of knowledge along the explicitness dimension. An associated KM activity (create, organize, formalize, share, apply, refine) further characterizes each knowledge flow taking place.

5.1.3 Ba: Shared Context for Knowledge Creation

According to (Nonaka and Konno 1998; Nonaka, Toyama, et al. 2000; von Krogh, Nonaka, et al. 2012), all activities of organizational knowledge creation have to take place in their appropriate context. This context is characterized by the (a) specifics of participants, and (b) their particular style of interaction (Nonaka, Toyama, et al. 2000). For each of the four knowledge conversion phases of the SECI model, Nonaka et al. (2000) introduce a dedicated type of *ba* that is characterized along the two dimensions of (a) the interaction type (*individual* vs. *collective*), and (b) the media primarily used during the interactions (*face-to-face* vs. *virtual*).

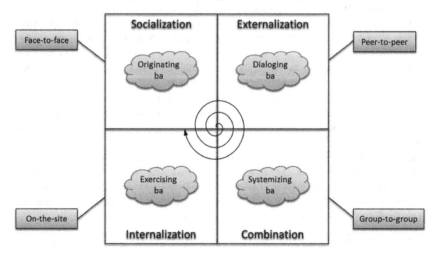

Figure 5.5:   Four types of *ba* within organizational knowledge creation, adapted from (Nonaka and Konno 1998) and (Nonaka, Toyama, et al. 2000)

In the following, the four types of *ba* depicted in Figure 5.5 will be briefly presented (Nonaka and Konno 1998; Nonaka, Toyama, et al. 2000):

(a) The *originating ba* stands for the context where the sharing of experiences and mental models between individuals takes place during the *socialization* phase. Typically, *face-to-face* interactions prevail in this context.

(b) The *dialoging ba* describes the context for the *externalization* phase where, in the manner of dialoging, the tacit knowledge of individuals is progressively made comprehensible to a larger group of people, and finally formalized in an explicit form. The interaction in the *dialoging ba* takes place in a *peer-to-peer* manner within a group providing a well-suited mix of knowledge for the particular problem.

(c) The *systemizing ba* provides the context for the *combination* phase and focuses on the handling of explicit knowledge primarily using IT, i.e. most interactions take place in a *virtual* manner between the various organizational groups.

(d) Within the *exercising ba*, explicit knowledge is converted into tacit knowledge through learning that takes place (a) during training activities or apprenticeships, and (b) the study of various artifacts of explicit knowledge (e.g. guidelines, reports, books, software).

5.1.4 Knowledge Resources

The knowledge base of an organization comprises both individual and collective knowledge resources (Romhardt 1998; Probst, Raub, et al. 2010). According to Nonaka et al. (2000), these knowledge resources assume a trifold role within the process of organizational knowledge creation:

(a) The knowledge resources supply the necessary knowledge as *input* to the SECI process.

(b) The SECI process creates new knowledge resources as its *output*.

(c) Likewise, knowledge resources are required within the four types of *ba*. The knowledge resources of the various participants and the combination of their knowledge profiles influence highly the success of the knowledge conversion activities.

Consequently, organizational knowledge resources are continuously altered as outcome of the SECI process and therefore have to be considered as *dynamic*. Nonaka et al. (2000) propose four distinct categories of knowledge resources that loosely map to the four phases of the SECI process (cf. Figure 5.6):

(a) *Experiential knowledge resources* comprise the shared tacit knowledge (e.g. skills, know-how) acquired through common experiences of individuals of an organization.

(b) *Conceptual knowledge resources* encompass the various artifacts of explicit knowledge remaining at a conceptual and rather abstract level, e.g. concepts, ideas, visions, and designs.

(c) *Systemic knowledge resources* comprise the well-documented, systematized, and easily transferrable artifacts of explicit knowledge, e.g. specifications, product models, guidelines, or patents.

(d) *Routine knowledge resources* encompass the shared tacit knowledge embodied by individuals and groups of an organization that is applied in the day-to-day operations.

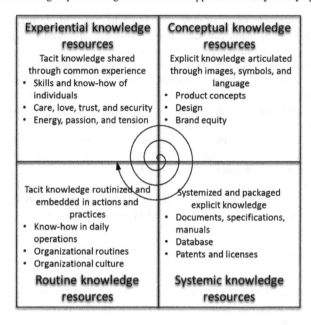

Figure 5.6:   Four categories of knowledge resources within organizational knowledge creation, adapted from (Nonaka, Toyama, et al. 2000)

5.2 Knowledge Management

The scientific field of *knowledge management* (KM) provides the contextual frame for analyzing and directing knowledge-intensive business processes. Based on the current research literature, the status of KM, its aims, and components will be highlighted in order to introduce the aspects of KM relevant for the present thesis. Finally, the current conception and practice of KM will be discussed.

5.2.1 Overview

Today, there is a large consensus on the perception of KM as a scientific field centered on the various activities, processes, and tools aimed at the management of the processes of knowledge creation, identification, recording, sharing, and application (Deng 2007; VDI 2009; Kebede 2010). Both Deng (2007) and Kebede (2010), have analyzed numerous KM definitions, and arrived by and large at similar definitions. Drawing from both authors, the present thesis will adopt the following definition for KM, which is outlined in the following:

Definition 5.2:   Knowledge management, adapted from (Deng 2007) and (Kebede 2010)

> Knowledge management comprises a set of strategies, methods, technologies, and tools for the purposeful and methodical management of knowledge creation, storage, sharing, and its application. It aims at realizing the full potential of knowledge in all kinds of knowledge-intensive business processes (e.g. decision making, problem solving) at all organizational levels (personal, group, organization, and inter-organization).

Research on KM as well as its practical implementation require an interdisciplinary collaboration of several disciplines (Romhardt 1998; Klabunde 2002) contributing to the targeted set of strategies, methods, and technologies. Figure 5.7 depicts the disciplines of information science, information systems, artificial intelligence, business administration, and work sciences as contributors to the field of KM with their potential inputs to KM (Romhardt 1998; Klabunde 2002).

Hansen et al. (1999) contrast two distinct strategies for KM, *personalization* and *codification*, a perspective that has been adopted by other KM researchers (Choi and Lee 2002; Larses and Adamsson 2004; McMahon, Lowe, et al. 2004) :

(a) *Codification* describes a KM strategy where knowledge is codified and subsequently stored in databases accessible by anyone in the organization. This *"people-to-documents"* approach suits organizations (i) selling standardized products, (ii) following a business strategy focusing on mature products, and (iii) with employees who tend to rely on explicit knowledge in their work process.

(b) *Personalization* describes a KM strategy where the knowledge is tightly associated with the person who created it and distributed mainly through personal communication. This *"person-to-person"* approach applies to organizations (i) selling customized products, (ii) following a business strategy based on product innovation, and (iii) with employees who tend to rely on tacit knowledge for problem solving.

Content Management
Meta Structuring
Knowledge Transfer

Learning
Further Education
Empowerment

Groupware
Workflow Management
Multimedia
Intranet/Internet

Knowledge Controlling
Intellectual Capital Report
Knowledge Strategy and Culture

Intelligent Agents
Expert Systems
Information Filtering

Figure 5.7: Main contributions to knowledge management, after (Klabunde 2002)

All knowledge-intensive business processes depend and interact with organizational knowledge, which therefore constitutes the point of reference for all KM efforts (Klabunde 2002). Accordingly, Figure 5.8 shows organizational knowledge in the focal point of the three levels of KM.

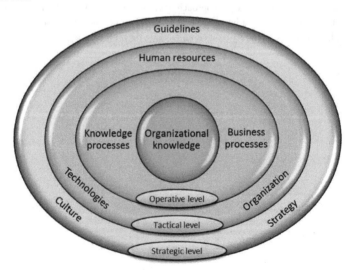

Figure 5.8: Functions and layers of knowledge management, according to (Klabunde 2002)

As part of any KM effort, dedicated knowledge processes have to support the interaction with organizational knowledge on these levels:

(a) At the *strategic level*, the knowledge processes have to focus on (i) deriving strategic knowledge goals from the strategic and normative business goals described within the business guidelines, given by the business culture, and the overall business strategy (Klabunde 2002), and (ii) on the identification of existing knowledge resources (Bodendorf 2006).

(b) At the *tactical level*, the realization of the identified knowledge goals has to be supported by a well-adjusted mix and coordinated effort of staff, an appropriate organizational structure, and information technologies (Klabunde 2002).

(c) At the *operative level*, the strategic knowledge goals have to be transformed into operative ones, directly influencing the day-to-day business processes (Bodendorf 2006). At the operative level, these business processes are complemented by six core activities of the KM process as depicted in Figure 5.9.

In 1995, the *geneva knowledge group*[27] introduced a phase model for the KM process that consisted of the five core activities *"Define knowledge goals"*, *"Develop knowledge"*, *"Share knowledge"*, *"Apply knowledge"*, and *"Assess knowledge"* (Romhardt 1998). Driven by critical feedback from industrial projects, many aspects of this initial model underwent improvements, until the final conceptual model for the components of KM emerged in 1996 (Romhardt 1998). Today, this model (cf. Figure 5.9) is widely accepted in the field of KM and cited and applied in many publications (Probst and Romhardt 1997; Romhardt 1998; Klabunde 2002; Bodendorf 2006; VDI 2009; Conrad 2010; Probst, Raub, et al. 2010).

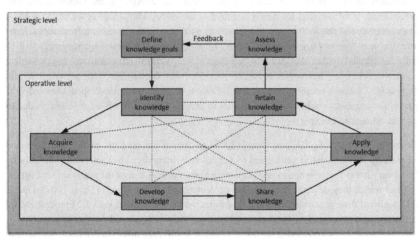

Figure 5.9: Core process of knowledge management, after (Probst, Raub, et al. 2010)

---

[27] Originally, *geneva knowledge group* was the designation of a KM research group at Geneva University. Later on, core people of this group (Büchel, Probst, Raub, Romhardt) founded a consulting company under the same name (Probst and Romhardt 1997).

Romhardt (1998) describes the conceptual model of KM primarily as a *heuristic* for the business practice providing (a) sufficient guidance for the KM process, and (b) enough degrees of freedom for interventions.

As there are entire publications available giving a very detailed presentation of the core activities of the conceptual model for KM (Romhardt 1998; Probst, Raub, et al. 2010), the KM process model will be described here only briefly based on (Romhardt 1998):

On the strategic level, an organization has to *define knowledge goals* that determine which aspects of its organizational knowledge will be extended. For this purpose, three different types of knowledge goals are required: *Normative* knowledge goals focus on developing the overall business culture toward knowledge awareness. *Strategic* knowledge goals define the competences of an organization required in the future. *Operative* knowledge goals focus on the implementation of KM in the day-to-day business processes. Finally, the associated controlling of the KM process assess to which degree the defined knowledge goals have been reached. For this purpose, dedicated methods for the measurement of the success of KM have to be provided (Housel and Bell 2001).

The first step on the operative level is dedicated to (a) *identify knowledge* already existing within an organization, and to (b) analyze and clarify the available external knowledge resources, e.g. patents, experts or innovative companies. In the next step, such externally available knowledge resources may be *acquired* e.g. by the licensing of patents, the hiring of experts or the acquisition of innovative companies, and subsequently integrated into the organizational knowledge. Moreover, the activity *"Develop knowledge"* focuses on internal knowledge creation through e.g. new capabilities, innovative products, and improved business processes. In a further step, the newly acquired or developed knowledge resources have to be made available to the individuals and groups of an organization benefiting from them. Even after the step *"Share knowledge"*, it is still not assured that the knowledge resources will finally be applied within the business processes. Here, various structural and psychological barriers have been overcome in order to *apply* the *knowledge*. The activity *"Retain knowledge"* focuses on the retention for the future of acquired or developed knowledge as part of the organizational knowledge in order to prevent losses of knowledge resources due to e.g. the retirement of experts.

Holsapple and Joshi (2003) and Deng (2007) compared the contents of the core activities from many KM frameworks proposed by the research literature. The conceptual models analyzed vary in (a) the designation and the description of the content of the activities, and (b) the level at which they describe a KM activity (elementary vs. high-level) (Deng 2007). Due to the identified divergences between the KM frameworks, Deng develops a KM process model dedicated to the domain of product development, which is centered on the perspectives of four distinct roles of knowledge agents or workers: *knowledge individual, knowledge analyst, knowledge engineer,* and *knowledge manager*. Deng (2007) defines for each of these roles specific KM process elements, e.g. *"Access knowledge"*, or *"Structure knowledge"*, which add up to 16 KM activity types altogether.

5.2.2 Critical Remarks

So far, the current conception of KM, its functions, layers, strategies, and core activities have been introduced. Over the years though, the presented conception of KM and likewise its practice have become the target of criticism. According to the analyses in (Levy 2009) and (Wolf, Rauhut, et al. 2009), quite a few authors consider the described practice of KM as dysfunctional and outdated. Consequently, some researchers label it as *"Knowledge Management 1.0"* and favor a new contextual and technological basis drawing from the *"Web 2.0"* technology and tool stack with special regards to social software (Wolf, Rauhut, et al. 2009). Bebensee et al. (2011) analyzed the impact of the *"Web 2.0"* applications on KM and (a) conclude that they may have a positive effect on certain KM aspects, and (b) see an even higher potential in a bottom-up approach for the adaptation of KM-relevant tools driven by the actual users.

In general, the authors criticize the following misconceptions within KM:

1. Due to the preference of the traditional western value system to explicit over tacit knowledge (Nonaka, Toyama, et al. 2000), in an inaccurate generalization, knowledge is misinterpreted as a *transferrable good*, which may be provided by a *central entity* (Wolf, Rauhut, et al. 2009). As described in its definition in section 3.2.1, only explicit knowledge may flow relatively freely between the different people and groups of an organization. Regarding the idea of the central provisioning of knowledge, von Krogh et al. (2012) analyzed the research literature on leadership in organizational knowledge creation and concluded that most publications believe in central leadership of this process, an idea they subsequently refute.

2. Wolf et al. (2009) describe an intrinsic motivational problem for knowledge sharing under the codification strategy, if the audience is not well defined. Although most people take pleasure in sharing their knowledge with others, they need a clearly defined context and audience providing feedback and motivation.

3. The whole idea that knowledge can be managed has been criticized by several authors (Wilson 2002; Wolf, Rauhut, et al. 2009). Sveiby (2001; cited in Wilson 2002) tellingly targets his critics upon the improper aim and naming of KM: *"I don't believe knowledge can be managed. Knowledge Management is a poor term, but we are stuck with it, I suppose. "Knowledge Focus" or "Knowledge Creation" (Nonaka) are better terms, because they describe a mindset, which sees knowledge as activity not an object. A is a human vision, not a technological one."*

Nissen (2006) focuses his points of criticism on the state of research on knowledge flows and derives from them topics for future research:

4. He characterizes KM largely as *descriptive theory* and remarks that the analysis and the shaping of knowledge flows require theories for explanation and prediction. Consequently, he raises the need to develop *theoretical and computational models of knowledge flows* and to stop focusing on technologies and flows of data and information.

5. Nissen demands that KM should *learn from physical sciences*, which have successfully dealt with dynamic phenomena for decades. He suggests that principles from physical sciences such as e.g. inertia, energy, and entropy might be applicable to the dynamics of knowledge as well. Moreover, he encourages KM researchers to adopt the same standards as their counterparts in physical sciences regarding research methods and the communication of results.

6. He prompts KM researchers to apply *mathematics* and *computational methods* within the analysis and simulation of knowledge flows.

6 Research Framework for the Analysis of the Knowledge Characteristics of Product Development

Designing is a process in which all sorts of things are done [...], but above all, it is a process of goal-directed reasoning.

Roozenburg and Eekels, 1995

The mechanical design process is a problem-solving process that transforms an ill-defined problem into a final product.

Ullman, 2003

[...] design consists in specifying an artifact (the artifact product), given requirements that indicate [...] one or more functions to be fulfilled, and needs and goals to be satisfied by the artifact, under certain conditions (expressed by constraints). At a cognitive level, this specification activity consists of constructing (generating, transforming, and evaluating) representations of the artifact [...].

Visser, 2006

As illustrated by the three quotes, the nature of the engineering design process can be perceived from rather different perspectives. In the first quote, Roozenburg and Eekels (1995) describe the engineering design process primarily as a reasoning process that uses different inference patterns to gain the required knowledge for decision making and the synthesis of design solutions. In contrast, Ullman (2003) comprehends the mechanical design process as a problem solving process for ill-structured problems, which are characterized by a lack of knowledge on the initial state, the goal state, and the steps required to reach the goal state. In the third quote, Visser (2006a) describes the design process mainly as sequence of transformation activities from one state of artifacts specifying the product to the next state. This multitude on opinions on the nature of the design process reflects the discussion of the various perspectives of cognitive psychology (reasoning, problem-solving, decision-making and judgment) introduced in chapter 4. Therefore, one of the topics dealt with in this chapter will be the most suitable approach provided by cognitive psychology for the description of (individual) cognitive activities in product development.

In addition, product development is these days typically conducted as an organizational activity. Therefore, the processes of knowledge creation at the individual and the organizational level have to be performed jointly in order to succeed in product development – Nonaka and

Takeuchi (1995) characterize this interdependency as the dichotomy of individual and organization within knowledge creation, where individuals create knowledge by their creativity, skills, and experience; knowledge that is subsequently amplified at the organizational levels. On the other hand, organizations enable and stimulate individual knowledge creation by providing a specific context (cf. *ba* in the model of organizational knowledge creation) that is characterized for instance by a specific level of standards, a certain degree of autonomy, and mutual trust between participants (Nonaka, Toyama, et al. 2000).

This sophisticated interplay of individual and organizational knowledge creation activities gives a clear indication that product development belongs to the category of knowledge-intensive business processes (Szykman, Racz, et al. 2000; Ameri and Dutta 2005; Deng 2007) that is characterized by the following attributes (Gronau, Müller, et al. 2004):

(a) High contribution of knowledge to the added value of the process
(b) Business processes consist of many creative parts
(c) Strong emphasis on communication
(d) Applied knowledge may have a short life-time; nevertheless the build-up of new knowledge is time and resource intensive

In contrast to conventional, i.e. rather rigid and predictable business processes encountered in many fields (e.g. manufacturing, administration, controlling), the product development process usually begins with a lack of knowledge about the end product and sometimes even the development approach to be adopted[28]. Whereas design methodology provides support to cope with the latter problem, the designers' skills, experiences, creativity, and ability to learn are the key factors to fill the gaps in knowledge during the development process.

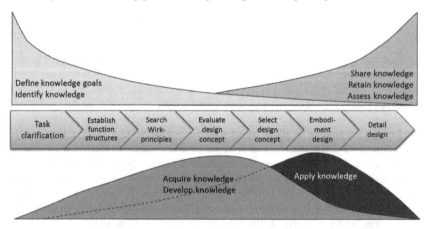

Figure 6.1:   Distribution of knowledge activities in the product development process, adapted from (Kaiser, Conrad, et al. 2008)

---

[28] This illustrates the ill-structured nature of the underlying design problems. In contrast to well-structured problems, ill-structured problems cannot be solved by following a known procedure.

Figure 6.1 depicts the typical distribution of KM activities focusing on knowledge specific to the end product and the development approach throughout the product development process. At the very beginning of the process, the dominant activities aim at the definition of required knowledge resources and the identification of knowledge already existing in the organization. The conceptual design phase focuses in a broad sense on *learning*, i.e. on the acquisition of relevant external knowledge resources and the creation of internal knowledge. It is not until the end of the conceptual design phase and during subsequent phases, that the actual application of acquired knowledge plays the dominant role.

Vajna (2006) characterizes engineering processes as *"dynamic, creative, chaotic"* and non-predictable. For such processes, the term *"workflow"* does not truly apply, as it characterizes primarily processes with reproducible working steps as encountered in manufacturing, controlling, and administration. The dynamic nature of product development processes is more appropriately reflected by the terms *"process"* and *"process element"* designating the different levels of engineering activities in product development (Vajna 2006).

Due to the creative and dynamic nature of product development, such process elements[29] may use and coincidentally alter the associated knowledge resources. Therefore, the intended analysis of knowledge characteristics in product development has to reflect appropriately such dynamics on the knowledge level. Overall, the intended research framework for the analysis of the product development process should cover the following topics:

(a) Knowledge conversions conducted within each process element
(b) Information and knowledge objects generated by these knowledge conversion activities
(c) Product development artifacts containing embedded knowledge from the conducted knowledge conversions
(d) Individual and organizational activities within each process element

Section 1.1 assesses which approaches of cognitive psychology are suitable for the description of (individual) cognitive activities in product development. Moreover, it lays out a view on the product development process as interplay of individual and organizational knowledge creation activities. Based on this perception, it introduces the descriptive model of knowledge creation in interdisciplinary product development. Next, section 6.2 provides an introduction on the various taxonomies for engineering knowledge within the product development domain. In section 6.3, the relevant requirements for the analysis of knowledge resources, knowledge contexts, and knowledge flows in product development will be gathered. Subsequently, section 6.4 presents the status quo on approaches for the analysis of knowledge characteristics and evaluates their ability to cover the collected requirements. Finally, section 6.5 will introduce the synthetized research framework aiming at the analysis of knowledge characteristics in product development and evaluate the achieved requirements coverage.

---

[29] According to (Freisleben 2001), a process element describes one activity, operation or one or multiple working steps. It is started by one or multiple events and ends in one or multiple events.

6.1 An Descriptive Model of Knowledge Creation in Interdisciplinary Product Development

Chapter 4 presented the multitude of approaches of cognitive psychology for the description of (individual) cognitive activities. It deferred the decision for selecting a particular approach for the description of cognitive activities for product development to the present section, as the selection of such approach needs to reflect the specifics of the cognitive activities conducted in the given context. In addition, the present section lays out a view on the product development process as interplay of individual and organizational knowledge creation activities. Based on this perception, it introduces the descriptive model of knowledge creation in interdisciplinary product development.

6.1.1 Individual Knowledge Creation within Product Development

The three approaches presented at the beginning of the chapter (reasoning, problem-solving, and construction of representations) comprehend individual knowledge creation in product development as *individual cognitive processes*. Some authors, however, report on cases that are best explained by *"socially distributed cognition"* among large groups: In one case, Knorr Cetina (1999) describes the epistemic culture within research on high-energy physics where several hundreds of people may work on different aspects of the same experiment. Within these self-organized groups, a communitarian culture prevails that considers it as impossible to determine the contribution of an individual to the overall success. Consequently, all contributors are mentioned equally in the resulting scientific papers. For this very distinct context, Knorr Cetina (1999) merely hypothesizes about *"distributed cognition"* taking place between the many different roles involved in the preparation and execution of these experiments.

Although *"socially distributed cognition"* appears as an interesting approach to explain social aspects of knowledge creation in product development, it presupposes acknowledgement that social groups possess the characteristics of cognitive entities. Cook and Yanow (1993) analyzed the question whether the cognitive perspective can be applied for explaining knowledge creation in a social context and uncovered three substantial problems that impede a simple transfer of the cognition-based model to organizations. Therefore, the present thesis adheres to the traditional view of cognition residing inside a person and adopts the following definition of individual knowledge creation:

Definition 6.1:   Individual knowledge creation

Individual knowledge creation comprises all kinds of cognitive activities (e.g. reasoning, creative thinking, learning) leading to an extension of an individual's knowledge. New knowledge is created through interactions between individual and environment and through new combinations of existing knowledge.

6.1.1.1 Product Development as Reasoning Process

In the following, the first of the three approaches presented at the beginning of the chapter will be introduced and discussed. This approach describes the engineering design process primarily as a reasoning process.

Overall, deduction and induction are the two modes of reasoning known and applied in many fields of science. Deduction represents a form of analytical reasoning that allows drawing conclusions from a given law and a particular fact. The conclusion is obtained in the form of a particular fact as indicated by Table 6.1. In contrast to other forms of reasoning, deduction is certain but it does not generate new knowledge in the forms of laws, rules, or theories. Table 6.1 gives an example of deductive reasoning for a window-lifting system with pinch protection. Furthermore, it describes the application of deduction within the context of simulation, where the behavior of the product under development is deduced from the simulation results (Roozenburg and Eekels 1995).

In contrast, induction allows establishing a hypothesis for a new law or a new theory from a number of observations pointing in the same direction. In contrast to deduction, the conclusion drawn by induction is only probable but not certain. Therefore, the hypothesized law or theory requires justification by deduction for a large number of cases and contexts. Nevertheless, the knowledge obtained remains empirical and should only be applied in the contexts where it has been sufficiently justified. Although inductive reasoning is typically encountered in empirical sciences, it plays also a certain role within product development, where it is required to synthetize generalizations from observations. Moreover, Lu and Liu (2012) describe inductive reasoning as the key mode of reasoning used in design evaluation, where experiments are carried out to determine whether the tested design iteration satisfies the functional requirements. Table 6.1 gives an example of inductive reasoning applied within the evaluation of two sensor concepts for a window-lifting system with pinch protection.

As already mentioned, deduction does not generate new knowledge in the forms of laws, rules, or theories. Likewise, Schurz (2008) points out similar limitations of inductions that *"cannot introduce new concepts or conceptual models"*. Consequently, both deduction and induction cannot sufficiently explain the creative synthesis activities within product development, where innovative solutions are conceived mainly driven by intuition and experience.

Therefore, C.S. Peirce introduced abduction as the mode of inference able to introduce new ideas in science, whereas deduction and induction cannot originate any ideas (Peirce 1903, CP 5.145). For this reason, abduction is recognized by many authors as the decisive mode for productive reasoning in engineering design (Takeda, Veerkamp, et al. 1990; Roozenburg and Eekels 1995; Tomiyama, Yoshioka, et al. 2002; Lossack 2006; Eder and Hosnedl 2008). Abduction focuses on establishing a hypothesis that allows an explanation of encountered evidence. It employs backward reasoning on known theories or laws to find the best of all possible explanations.

Table 6.1:   Traits of deduction and induction and their application in product development

	Analytical reasoning		Synthetical reasoning	
	Deduction		Induction	
Known	$\forall x(C(x) \rightarrow E(x))$ $C(a)$	Law or theory Singular fact	$C(a) \rightarrow E(a)$ $C(b) \rightarrow E(b)$ $C(c) \rightarrow E(c)$	Singular observation Singular observation Singular observation
Conclusion	$E(a)$	Singular fact	$\forall x(C(x) \rightarrow E(x))$	New law or theory
Example	If x contains a pinch protection then the passenger adjacent to x is safe from accidental clamping		The x using two hall sensors measures the window position more precisely than an x solely based on software sensors.	
			The x using two hall sensors measures the window velocity more precisely than an x solely based on software sensors.	
	x contains a pinch protection		The x using two hall sensors measures the moving direction of the window more precisely than an x solely based on software sensors.	
	The passenger adjacent to x is safe from accidental clamping		An x using two hall sensors is more reliable than an x solely based on software sensors.	
Application in product development	In simulation, deduction is applied to establish a prediction for the expected behavior of the product under development. In a preparatory step, a model is established that serves as logical system for deductions. Based on the model, experiments are carried out that lead to results. From these experimental results, the behavior of the product under development is deduced. (Roozenburg and Eekels 1995)		Usually, inductive reasoning is applied when experimenting e.g. in the context of design evaluation. Here, experiments are carried out to determine whether the tested prototype satisfies the functional requirements. (Lu and Liu 2012)	

Abductive reasoning is commonly encountered when (a) establishing a diagnosis in medicine or in technical defect tracking, (b) searching for the best explanation of given evidence in criminal investigations, and (c) establishing new laws, theories or theoretical concepts in scientific research. A rather simple case in the context of criminal investigations describes the typical pattern of abductive reasoning:

Used background knowledge:	If x enters a location, x's shoes leave characteristic footprints.
Encountered evidence:	Footprints were found at the crime scene and identified as belonging to x's shoes.
Hypothesis for best explanation:	x has been at the crime scene and can therefore be considered as a suspect.

This example illustrates that abduction leads to a merely possible hypothesis that requires further analysis and justification to become accepted as a probable explanation. Altogether, abduction, deduction, and induction form the *cycle of inquiry*, where in the first step abduction develops a hypothesis. In the second step, the consequences of this hypothesis are further analyzed by deduction. In the third step, these explicitly stated consequences of the hypothesis are tested by induction (Staat 1993).

Roozenburg and Eekels (1995) distinguish two fundamental patterns of abduction designated as *explanatory* abduction and *innoduction* influenced by the interpretation of Habermas (1973) given for abduction. In their understanding, *innoduction* allows inferral or rather projection from an intended result (e.g. the intended function of a product given by a functional specification) to both the solution (e.g. a design solution realizing the intended function) and an accompanying theory that explains the realization. Schurz (1995) denotes the schema applied by *innoduction* as *speculative* abduction and criticizes this schema as unusable. In a later publication, Schurz (2008) develops a differentiation of scientifically fruitful abductions from *speculative* abductions by introducing the criterion of causal unification, where he positions the *Reichenbach principle* as the leading approach for causal unification. When encountering several interrelated phenomena, this principle allows the conclusion that either (a) one phenomena is the cause of the other one, or (b) these phenomena must have a common cause that explains all of them (Schurz 2008).

Overall, the research community in product development has reacted rather reluctantly to the concept of *innoduction* and it has only been adopted by a few authors (Nemeth 2004; Eder and Hosnedl 2008), whereas the importance of *abduction* has been widely recognized. Consequently, the present thesis will exclude *speculative* abduction and adopt the elaborated systematics of abduction proposed by (Schurz 2008). In the first step, Schurz (2008) designates the two previously discussed patterns of abduction as *creative* abduction that introduces new concepts and *selective* abduction that focuses on choosing the best explanation among a set of possible explanations. In the second step, Schurz (2008) introduces a further classification for patterns of abduction based on the following three interrelated criteria:

(a) The kind of *hypothesis* that is abduced, where new facts, new laws, or new theories in conjunction with new concepts are introduced.
(b) The kind of *evidence* (e.g. empirical facts, empirical laws, and general empirical phenomena) which the abduction aims to explain
(c) The *beliefs* or *cognitive mechanisms* driving the abduction, where known laws or theories, theoretical background knowledge, speculation and causal unification can possibly be applied.

From the resulting categories of abductions, Table 6.2 explains the characteristics of the three categories of abduction relevant in product development. Firstly, *factual* abduction establishes a hypothesis in the form of a singular fact for given evidence as a singular fact. It employs backward reasoning on known theories or laws to find the abduced hypotheses. Tomiyama, Takeda, et al. (2003) remark that factual abduction remains, however, within the limits of the given background knowledge. It cannot explain the specific traits of creative and innovative design.

Table 6.2:   Types of abduction and their application in product development

	Synthetical reasoning				
	Abduction				
	Factual abduction		Law-abduction		2nd order existential abduction
Beliefs driving abduction Evidence	$\forall x(C(x){\rightarrow}E(x))$ $E(a)$	Known law or theory Singular fact	$\forall x(C(x){\rightarrow}E(x))$ $\forall x(F(x){\rightarrow}E(x))$	Background law Empirical law	Theoretical background knowledge Empirical phenomena (laws)
Conclusion, i.e. abduced hypothesis	$C(a)$	Singular fact	$\forall x(F(x){\rightarrow}C(x))$	Law	Law or theory with new concepts
Example	If x contains a pinch protection then the passenger adjacent to x is safe from accidental clamping		Within x, the velocity of the window is not measured directly. It is calculated as the first derivative of the observed position.		Knowledge on the micro-structure of materials, especially in combination with the electronic energy band model
	The passenger adjacent to x should be safe from accidental clamping		The x using two hall sensors measures the window velocity more precisely than an x solely based on software sensors.		Certain kind of materials (e.g. iron, tin, or copper) expose similar characteristics such as high conductivity of heat and electricity, hardness, and elasticity.
	x must contain a pinch protection		The x using two hall sensors measures the window position more precisely than an x solely based on software sensors.		The common cause of the observed commonalties is the metallic character of these materials.
Application in product development	In synthesis, factual abduction (in particular first- order existential abduction) is able to generate an entity fulfilling the requirements from a known set of solutions. (Tomiyama, Takeda, et al. 2003)		In synthesis, law-abduction is applied when experimenting e.g. with new solution concepts. Through experiments, empirical laws are observed. For their explanation a hypothesis in the form of a new law is developed.		In the synthesis process for interdisciplinary products, analogical abduction can be applied to establish physics-based multi-domain models. (Tomiyama, Takeda, et al. 2003)

Table 6.2 gives an example of *factual* abduction for a window-lifting system where a given requirement is employed to establish a hypothesis on the characteristics of a product component. Furthermore, it describes the application of *factual abduction* within synthesis. Secondly, *law*-abduction creates a theoretical hypothesis (new law) for given evidence in the form of an empirical law by applying background laws. Table 6.2 gives an example of *law*-abduction applied within the synthesis of a sensor concept for window-lifting system with pinch protec-

tion. Furthermore, it describes the application of *law abduction* in the frame of experimenting with new solution concepts in synthesis.

Schurz (2008) emphasizes the *selective* character of the two kinds of abduction introduced so far, i.e. they focus on choosing the best candidate among a set of possible explanations. In contrast, *second-order-existential* abduction is merely *creative*, i.e. it is able to introduce new theoretical concepts. *Analogical* abductions and *common cause* abductions can be mentioned as typical examples of *second-order-existential* abductions. *Analogical* abduction employs analogies between the evidence to be explained and known theories to establish a new theory associated with analogical concepts. Tomiyama, Takeda, et al. (2003) describe the process of establishing analogies for physical models from one to another domain (mechanical and electrical) as an example of *analogical* abduction (cf. Table 6.2). *Common cause* abduction allows finding a common cause for several interrelated phenomena. It applies the knowledge about those phenomena in the search for a unified cause of these phenomena. Schurz (2008) gives an example of *common cause* abduction that reasons on the common cause of the common phenomena (e.g. high conductivity of heat and electricity, hardness, elasticity) of materials such as iron, tin, or copper. The knowledge applied for the abductive inference is the electronic energy band model that allows characterization of all these materials as *metal*.

In conclusion, C.S. Peirce tellingly illustrates the different degrees of certainty of the three aforementioned modes of reasoning (Peirce 1903, CP 5.171): *"Deduction proves that something must be; Induction shows that something actually is operative; Abduction merely suggests that something may be."*

Although the presented perspective on product development as reasoning process provides sound explanations for the main characteristics of the cognitive activities conducted during different phases of the product development process, there is empirical and theoretical evidence contradicting this approach:

(a) According to the previously presented dual-processing theories, intuitive elements and heuristics play an important role especially in creative thinking. As presented in section 4.2, creative thinking depends on both hypothetical reasoning and expert intuition.

(b) Empirical studies on design processes reveal only rare application of deductive inferences (Visser 2006b).

Comparable to the role of logical reasoning in cognitive psychology as *normative* theory, the description of engineering design by the means of reasoning should likewise be understood as a *normative* approach. It describes the conducted cognitive activities only under "ideal" conditions (ideal knowledge, no need for Type 1 processing due to unlimited cognitive resources conducting Type 2 processing). Although it does not sufficiently reflect the multitude of cognitive operations conducted during engineering design by human agents, it provides, however, an interesting approach for conceiving artificial agents in design support.

6.1.1.2 Product Development as Construction of Representations

From a cognitive viewpoint, Visser (2006b) distinguishes two approaches for describing design. The first approach understands design mainly as the cognitive activity of *problem solving*, whereby the design problem is typically considered as ill-structured. The second approach perceives design primarily as *situated action and cognition*. It comprehends design primarily as a making process, where the problems to be solved are constructed from the real-world settings. This perspective on the design process as reflective practice was proposed by Schön (1984): Firstly, designers *name* and *frame* a design problem, and secondly *move* to develop a solution for the given problem.

Visser (2006a) compares these approaches and concludes that both of them neglect essential cognitive aspects of the design process. Consequently, she proposes a new approach that comprehends design mainly as *construction of representations* (i.e. the various types of design models) and adopts at the same time certain elements of the previously presented approaches. Deng (2007) describes the closely related *state-process-resource model* that perceives the product development process as a sequence of process elements transforming the product under development from a given state to the subsequent one (cf. section 6.4.6). From the previously presented problem solving approach, Visser (2006a) adopts the idea of perceiving problems in terms of the representations constructed for these tasks. Likewise, from approaches focusing on situated action and cognition she endorses the perspective on design as an active and constructive process.

Drawing from the (Goel 1995) and other researchers, Visser (2006a) describes three types of cognitive activities on representations within her framework:

(a) *Generation* describes the construction of product representations starting mainly from mental models.

(b) *Transformation* comprises cognitive activities modifying an input representation R_i and leading to an output representation R_{i+1}.

(c) *Evaluation* represents cognitive activities for assessing how well a design solution described by representations fulfills the given requirements.

Within this framework of cognitive design, transformation activities exist in a large variety of forms. Overall, they can be characterized by the specifics of the transformation from R_i to R_{i+1} (Visser 2006a):

(a) *Duplication* generates a replicate of R_i.

(b) *Addition* augments the input representation R_i by new information and small modifications.

(c) *Detailing* decomposes R_i into multiple components.

(d) *Concretizing* transforms R_i into a more concrete (i.e. less abstract) form.

(e) *Modifying* comprises all other transformation activities of R_i besides detailing and concretizing leading into R_{i+1}.

(f) *Revolutionizing* transforms R_i by a creative leap into an alternative representation R_{i+1}.

Other sub-activities and operations play a supportive role for the mentioned types of cognitive activities on representations. Visser (2006a) cites here the examples of *interpretation, articulation, analysis, exploration, combining, hypothesizing,* and *justifying.*

Figure 6.2 visualizes the adopted understanding of the cognitive activities of an individual in product development that consists of:

(a) The *tripartite model of mind* introduced in section 4.3 for the description of the involved cognitive systems, the relationships between cognitive systems, and the relationships to knowledge.

(b) The various types of activities described by (Visser 2006a) that either *generate* an initial state of the product representation or *transform* an existing state to the subsequent one. In addition, different strategies for the *evaluation* of a reached state of design are employed.

(c) The various *representations* describe the product at a given state. The arrow leading from Representation$_i$ to Representation$_{i+1}$ represents the actual transformation of this artifact by an associated cognitive activity. In contrast, the arrow leading back from Representation$_{i+1}$ to Representation$_i$ indicates an evaluation activity that assesses the characteristics of the achieved solution by taking into account elements of the anterior representation (e.g. requirements, functions, or constraints).

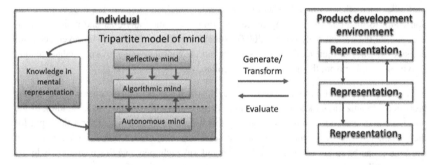

Figure 6.2: Integration of the tripartite model of mind (Stanovich 2009; Stanovich 2011) with the framework for cognitive design research (Visser 2006a) for describing the cognitive activities of an individual interacting with the product development environment

In summary, the view introduced for the product development process as a sequence of constructions of representations (i.e. models) matches well with the specific interest of the present thesis in these representations, as they contain embedded information and knowledge to be extracted and employed by the intended design support system. Moreover, the three types of cognitive activities on representations proposed by (Visser 2006a) and associated subactivities will be adopted for the analysis of the knowledge conversions in individual knowledge creation.

6.1.2 Organizational Knowledge Creation in Product Development

Nonaka and Takeuchi (1995) build their model of organizational knowledge creation (cf. section 5.1.1) on the foundations of the philosophical direction of *Pragmatism* and Japanese philosophy. Their model explains organizational knowledge creation through experiential learning and the continuous conversion process between tacit and explicit knowledge. When applied to product development, it remains, however, unclear whether this model can comprehensively describe the problem solving and decision-making processes in this domain. As a starting point for such clarification, the overall assumptions of the model will be examined.

In one instance, Nonaka and Takeuchi (1995) state that their model of knowledge creation is based on the *"critical assumption that human knowledge is created and expanded through social interaction between tacit and explicit knowledge"*. They further make clear that this process of knowledge creation is always a social process taking place between individuals but not within an individual. In this way, they explicitly deny the relevance of individual knowledge creation. In a different instance, however, they emphasize that *"knowledge is created only by individuals"* and that such knowledge is merely amplified and extended within the process of organizational knowledge creation to reach higher organizational levels (Nonaka and Takeuchi 1995).

Obviously, these two conceptions of the role of individual and collective knowledge creation contradict each other. Generally, theories of organizational learning[30] in the *Pragmatist* tradition claim to integrate the learning processes common to the individual and collective contexts in one theoretical body (Nobre 2007). Nonaka and Takeuchi (1995), however, seem indecisive on whether their model suffices to explain the complete range of knowledge creation activities spanning from the individual to the collective.

When studying the SECI process (cf. Figure 5.3), which serves as a conceptual model for the process of organizational knowledge creation, there is still only a vague understanding of the processes that lead to the initial acquisition of tacit knowledge and which form the input for the socialization activity. Due to the close association of the SECI model with experiential learning, it can be safely assumed that Nonaka and Takeuchi perceive this tacit knowledge as built-up while executing professional tasks. The characteristics of this emergence process, however, remain unexplained. In order to understand the characteristics of thinking activities and their interplay in product development, a more detailed modeling of this emergence process of individual knowledge seems necessary.

In later years, Nonaka, von Krogh, et al. (2006) clarified the focus of the model of organizational knowledge creation and considered individual knowledge creation as outside of their model. The following definition compiles the understanding adopted by the present thesis on organizational knowledge creation:

---

[30] Here, organizational learning is considered merely as a synonym for organizational knowledge creation.

Definition 6.2:   Organizational knowledge creation (Nonaka, von Krogh, et al. 2006)

> *"Organizational knowledge creation is the process of making available and amplifying knowledge created by individuals, as well as crystallizing and connecting it with an organization's knowledge system."*

6.1.3 Synthesis of the Descriptive Model for Knowledge Creation in Interdisciplinary Product Development

Each of the two introduced approaches (individual knowledge creation perceived as construction of representations vs. organizational knowledge creation) explains certain facets of the overall thinking and decision-making processes in product development. None of them alone, however, is able to capture and model (a) the complete range of sources of knowledge as well as (b) the interplay of individual and collective processes in product development. On the one hand, the framework on cognitive activities in design proposed by (Visser 2006a) is well suited to explain the patterns of knowledge application and creation within individuals (cf. Table 6.3). On the other hand, the model of organizational knowledge creation adopts a *sociocultural* perspective and neglects essential processes at the individual level. In addition to the different perspectives of the two approaches, Table 6.3 compiles the overall characteristics of these two models of knowledge creation.

In particular, this comparison emphasizes the different reach and scope of the two modes: While individual knowledge creation focuses on the extension of the individual competences in order to conduct or contribute to a certain portion of the product development process, organizational knowledge creation aims at extending the organizational competences in order to carry out product development as a whole. Within individual knowledge creation, the cognitive limits of an individual are not sufficient to cover the complete scope of required disciplines. Therefore, only organizational knowledge creation permits an appropriate coverage of the full scope of involved disciplines and interdisciplinary topics as well as gaining knowledge for all parts of the development process. Finally, both modes of knowledge creation tightly interact: While individual knowledge creation provides the knowledge that is subsequently amplified at the organizational levels by organizational knowledge creation, the context for individual knowledge creation is provided by organizational knowledge creation.

As stated at the beginning of the chapter, both modes of knowledge creation have to be performed jointly in order to succeed in product development that is today typically an organizational activity. Therefore, the two approaches need to be incorporated into an integrated descriptive model of knowledge creation. This descriptive model should allow for a comprehensive modeling of the product development process, i.e. it should be able to capture appropriately the interdisciplinary nature of "modern" product development.

Table 6.3:   Traits of individual and organizational knowledge creation in product development

	Individual knowledge creation	Organizational knowledge creation
Perspective	*Cognitive* perspective: Individual as cognitive entity interacting with product development artifacts and environment	*Socio-cultural* perspective: Group as social entity constituted of individuals but governed by an organizational culture
Type of learning	Implicit and explicit learning processes	Focus on experiential learning leading to a built-up of tacit knowledge while executing professional tasks
Scope of knowledge creation	Extending individual knowledge usually within disciplinary limits	Extending knowledge of the organization at disciplinary and interdisciplinary levels
Reach of knowledge creation	Gaining knowledge for a specific portion of the product development process	Gaining knowledge for the overall product development process
Interaction patterns	Between individual, product development artifacts, and environment, in particular: - Generation of product representation: Individual and product development artifact - Transformation of product representation: Individual and product development artifact - Evaluation of product representation: Individual and product development artifact - Externalization: Individual and environment - Internalization: Individual and environment	- Socialization: Between individuals - Externalization: Within a group - Combination: Within a group, but usually virtual interactions - Internalization: Between individuals (training)
Cross-actions		Enables and stimulates individual ⇐ knowledge creation by providing a specific context
	Provides knowledge to be amplified at the organizational levels ⇒	

At first sight, two integration approaches seem obvious: The first approach aims at extending the use of the cognitive model of individual knowledge creation toward the organizational side. As discussed in section 6.1.1, such transfer of a cognition-based model to organizations is impeded by substantial problems. The second approach proposes employing the socio-cultural model of organizational knowledge creation to individuals. This line of thought, however, has already been dismissed in the previous section because it does not permit modeling the emergence of individual knowledge at a sufficient level of detail.

Therefore, a third integration approach is proposed that aims at integrating the two models at the level of their common conceptual elements. Both models share several concepts that can be employed as connection and integration points of them:

(a) *Knowledge conversion* takes place in both models: Individual knowledge is transformed by cognitive activities for the construction of representations. In addition, individual knowledge can be extended by learning (internalization) and the acquired knowledge can be externalized. At the organizational level, the conversion is performed by the four conversion modes of the SECI-model.

(b) Consequently, different types of *knowledge objects* serve as input for these conversion processes and will be altered as the result of the conversion. The tacit and explicit knowledge of an individual constitute the applied knowledge resources within individual knowledge creation, whereas the knowledge resources of organizational knowledge creation are located at the different organizational levels (e.g. individual, team, organization, etc.). Both models equally employ individual knowledge resources and explicit knowledge resources, which will therefore be considered as the integrating knowledge resources of the two processes.

(c) The *actors* of individual knowledge creation will re-appear on the organizational side either directly or as members of teams and other organizational units.

(d) To a first approximation, the actor in individual knowledge creation interacts only with its environment. Nevertheless, individual knowledge creation is strongly influenced by the organization through (i) goals defined by managers of the organization, (ii) prevailing organizational standards that largely determine the quality of the acquired knowledge, and (iii) the implicit extension of individual knowledge resources by the experiences gained when an individual executes tasks as part of a team.

From the point of view of individual knowledge creation, Figure 6.3 lays out the interaction model between the environment and the individual, who applies cognitive abilities and knowledge resources. When interacting with the environment, the knowledge resources support an individual's cognitive abilities to perform purposeful actions upon the environment. Subsequently, the results caused by these actions are observed and compared to the anticipations. The cognitive activities conducted by the various cognitive systems in combination with the existing knowledge resources infer new insights from these results. The gained insights lead to an extension of the individual's knowledge resources. In addition, the interaction model captures the interplay of the individual with the context of the organization: Directives issued by the organization's management and organizational standards govern the individual. Moreover, the individual can externalize certain of its tacit knowledge resources as well as internalize explicit knowledge resources from the organizational context.

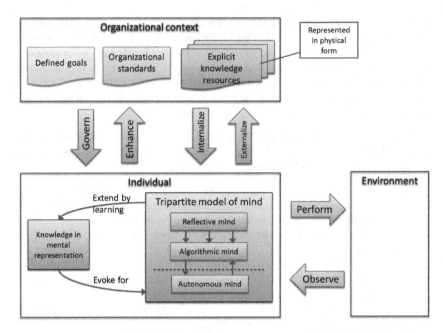

Figure 6.3:   Descriptive model of individual knowledge creation: Interactions of an individual with the environment and the context of the organization

For product development, this interaction model can be further specialized concerning (a) the characteristics of the concepts involved in the interactions and (b) the specifics of the encountered interactions (cf. Figure 6.4). Usually, the stakeholders of the product development organization compile the problem definition and associated goals for a new product into a requirements specification, which provides the input for a subsequent analysis activity of the individual leading to the synthesis of the product's function structure. Drawing from Visser (2006a), the activities of generation, transformation and evaluation of product representations (cf. section 6.1.1.2) are considered as the dominating cognitive activities within engineering design. Due to the synthesis character of the generation and transformation activities, the product representations contain information and knowledge embedded by these cognitive activities.

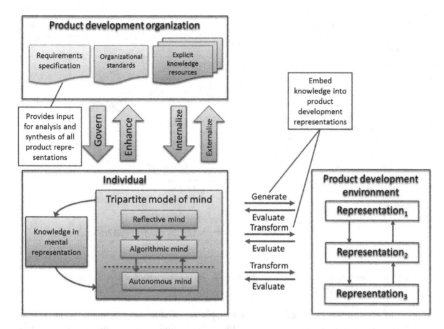

Figure 6.4: Descriptive model of individual knowledge creation in product development: Interactions between the individual with the product development environment and the product development organization

So far, the aforementioned descriptive models remain within the limits of the cognitive processes and competences of an individual and do not permit the capture of interactions between individuals, teams and other organizational units as occurring in organizational knowledge creation. In order to capture both modes of knowledge creation within an integrated descriptive model of knowledge creation, the aforementioned models have therefore to be extended by the interactions between these social entities. In particular, all four knowledge conversion modes of the SECI-model (cf. section 5.1.1) have to be included. In addition to externalization and internalization conducted as individual activities, in organizational knowledge creation both conversion modes are performed as social activities. Figure 6.5 depicts the constituents of the integrated descriptive model of knowledge creation and their interactions. Here, filled arrows indicate interactions involving an individual only and social interactions at the different organizational levels are shown by arrows without filling.

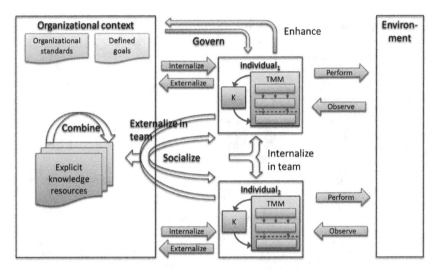

Figure 6.5:   Integrated descriptive model of knowledge creation covering individual and organizational
              knowledge creation

In order to account for the specifics of interdisciplinary product development, the integrated descriptive model can be further refined. When developing interdisciplinary products, the individuals involved possess knowledge profiles[31] specialized to their respective disciplines. This results in a fragmentation of the knowledge profiles of the involved actors that limits the social knowledge conversion modes to the few participants possessing the required depth of knowledge for the respective activity. In addition, the explicit knowledge resources within the organizational context and product artifacts in the interdisciplinary product development environment are fragmented into system-level and disciplinary components (cf. Figure 6.6). Consequently, an individual interacts only with the subset of them he/she is knowledgeable.

Altogether, the integrated descriptive model of knowledge creation allows describing the fragmented and heterogeneous activities of knowledge creation, its application, sharing, externalization, and internalization in interdisciplinary product development. It is therefore able to capture the complete range of sources of knowledge in interdisciplinary product development. In particular, it takes into account activities conducted by individuals only and social interactions at the different organizational levels and captures their interplay. Overall, the laid-out model encompasses both the *cognitivist* aspects of knowledge creation as well as the *social* and *cultural* dimension of knowledge creation.

---

[31] The knowledge profile captures the width and depth of knowledge from an individual, a team, or an organization. It is typically depicted in a two-dimensional diagram, where the horizontal axis captures the width of the knowledge distributed over the various knowledge domains, whereas the vertical axis symbolizes the depth of knowledge, i.e. the achieved level of expertise. The knowledge profile typically visualizes an increased depth of knowledge toward the bottom. This way, it depicts a knowledge profile with large background knowledge in many domains and deepened knowledge in one area as T shape.

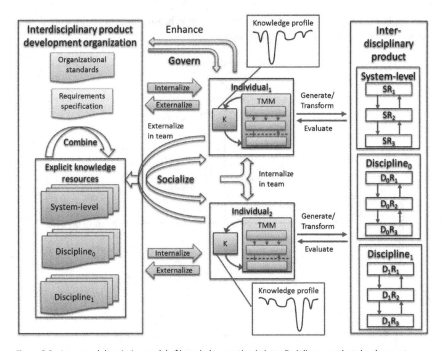

Figure 6.6: Integrated descriptive model of knowledge creation in interdisciplinary product development

The decomposition of a typical product development activity (e.g. spatial partitioning) down to the level of individual and organizational knowledge conversion modes, however, appears to be quite labor intensive. Such low-level analyses, though, can uncover the specific knowledge characteristics of the process elements appearing in product development processes.

6.2 Taxonomies of Knowledge in Product Development

Section 3.2 departed from the line of thought that no single, universally valid taxonomy of knowledge exists and that in many situations a rather pragmatic approach for the classification of knowledge is adopted. In the following, the research literature on product development is therefore briefly analyzed concerning the proposed approaches to the categorization of knowledge in this domain. Hubka and Eder (1992) identified a large number of knowledge domains as contributors to design engineering and emphasize that it is possible to codify and structure these various islands of knowledge into the form of taxonomies, models, and surveys. To a first approximation, they classify knowledge in design engineering into the categories of (a) knowledge on *design objects* and (b) knowledge about the *design process*, a classification shared by many authors (Hubka and Eder 1992; Horváth 2004; Ahmed, Hacker, et al. 2005; Eder and Hosnedl 2008). Complementarily, they use the dichotomy of *declarative* and *procedural* knowledge as a second dimension to distinguish (c) *descriptive* statements related to theories and (d) *prescriptive* statements with a practical focus. As depicted in Figure 6.7, they apply the resulting 2x2 matrix to provide a comprehensive map of engineering design science (Hubka and Eder 1992; Eder and Hosnedl 2008).

Figure 6.7:   Map of engineering design science, adapted from (Eder and Hosnedl 2008)

In the first dimension of their knowledge taxonomy (cf. Figure 6.8), Venselaar, van der Hoop, et al. (1987) describe the three classes of (a) domain-specific basic knowledge, (b) domain-specific design knowledge and (c) general process-aimed knowledge focusing on the methods and methodologies of the design process. For the second dimension, they also adopt the dichotomy of *declarative* and *procedural* knowledge, which is complemented by the supplementary categories of *situational* and *strategic* knowledge (cf. section 3.2.3).

Kim, Hwang, et al. (2003) propose a three-dimensional categorization (cf. Figure 6.9) that selects the knowledge scope as first dimension comprising general, domain and system knowledge. The second dimension focuses on the different types of knowledge, for instance procedural and catalog knowledge, where the latter captures the ability to locate relevant information sources. Finally, the third dimension describes the different usage levels of knowledge (factual, tactical, and strategic).

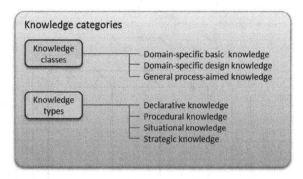

Figure 6.8: Categories of the knowledge taxonomy proposed by (Venselaar, van der Hoop, et al. 1987)

In the context of the aerospace industry, Ahmed, Hacker, et al. (2005) conducted an empirical study that verified the relevance of 24 knowledge categories established by a previous research project. The definition of the knowledge categories in the areas of (a) knowledge of the products, (b) process knowledge, and (c) managerial knowledge followed largely a pragmatic approach collecting the prevailing types of knowledge within the particular industry and for the considered product.

In the context of the process-based analysis of product development by the state-process-resource model, Deng (2007) proposes a two-layer model consisting of the level of (a) knowledge categorization (concepts, practice, knowledge about experts) and the level of (b) knowledge subjects that comprise e.g. the product, processes, methods, and tools.

Initially, Roth, Binz, et al. (2010) introduce a categorization of knowledge in product development based on multiple dimensions comprising (a) the type of knowledge with regards to the domain and various other criteria, (b) specific attributes (e.g. individual/collective, intern/extern) and additional criteria. Subsequently, they solely utilize the knowledge type to indicate the characteristic forms of knowledge within the phases of the product development process. Difficulties in the application of this knowledge taxonomy may arise from (a) the high number of 14 knowledge types and (b) the parent-child relationships between knowledge types.

Based on the current research literature, relevant approaches for the systematization of knowledge in product development have been presented. The initial statement that there is no single, universally valid taxonomy of knowledge holds also true in product development. The intended analysis of knowledge resources, however, requires means for the categorization of knowledge based on a specific taxonomy.

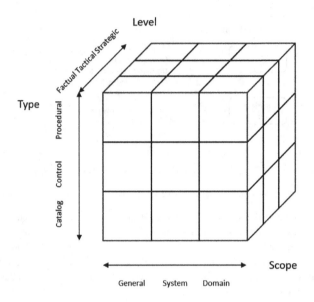

Figure 6.9:   Three-dimensional knowledge categorization proposed by (Kim, Hwang, et al. 2003)

Among the presented taxonomies, though, certain classification criteria seem to be widely accepted as they appear in several of them. Consequently, they should be considered as candidates for the knowledge taxonomy of the present thesis:

(a) The *subject* criterion comprises the categories of knowledge on the product and knowledge on development processes (Hubka and Eder 1992; Horváth 2004; Ahmed, Hacker, et al. 2005; Eder and Hosnedl 2008), managerial knowledge (Ahmed, Hacker, et al. 2005; Roth, Binz, et al. 2010), and supplementary subjects (e.g. methods and tools) (Deng 2007).

(b) The *domain* criterion encompasses the various knowledge domains (e.g. general, domain, system) (Venselaar, van der Hoop, et al. 1987; Kim, Hwang, et al. 2003).

(c) The dichotomy of *declarative* vs. *procedural* knowledge (Venselaar, van der Hoop, et al. 1987; Hubka and Eder 1992; Deng 2007; Eder and Hosnedl 2008) complemented by the categories of *situational* and *strategic* knowledge (Venselaar, van der Hoop, et al. 1987).

The aforementioned support for organizational knowledge creation necessitates the inclusion of its two dimensions (a) *explicitness* and (b) *organizational reach* within the knowledge taxonomy of this thesis that is depicted in Figure 6.10. In addition to the criteria themselves, it contains possible values for each criterion. As described in section 3.2.4, the adoption of model of organizational knowledge creation by the present thesis implies also the adherence to its vocabulary of concepts. Therefore, *explicit* knowledge will be considered synonymous to *declarative* knowledge as well as *tacit* knowledge synonymous to *procedural* knowledge.

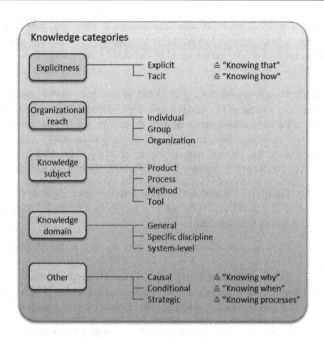

Figure 6.10: Criteria forming the knowledge taxonomy in product development

6.3 Requirements for the Analysis of Knowledge Characteristics within Product Development

This section will establish a comprehensive collection of requirements for the analysis of knowledge characteristics within product development based on the descriptive model of knowledge creation in interdisciplinary product development introduced in section 6.1.3. Primarily, these requirements will serve as evaluation criteria for existing analysis approaches. In addition, the gathered requirements will permit the identification of the potential for improvements within these approaches.

Requirement 6.1: In general, the analysis on the knowledge characteristics needs to be tightly integrated with the various activities comprising the product development process. Coincidentally, these business activities constitute the context for knowledge creation and associated knowledge utilization and transformation (Newman 2003). Therefore, the analysis on the knowledge level should use the various activities of the associated product development process as *integration* and *anchor points*.

Requirement 6.2: By compiling the opinions of several authors, Pogorzelska (2009) emphasizes that the knowledge flows of a given business process do not run along the lines of the business activities. Instead, the knowledge flows may run back and forth between the various activities and involved actors. Consequently, the analysis method should permit the modeling of *knowledge flows not aligned* with the various activities of the associated product development process.

As a whole, the descriptive model of knowledge creation in interdisciplinary product development provides appropriate guidance on the various aspects of knowledge (knowledge resources, knowledge conversion, and actors) to be included by the research framework. Therefore, each of the aforementioned knowledge traits will be covered by a separate requirement.

Requirement 6.3: According to Nonaka et al. (2000), *knowledge resources* assume three distinct roles within the process of knowledge creation (cf. section 5.1.4), where (a) knowledge serves as input to the knowledge creation activity, (b) new knowledge is created as output of an activity, and (c) knowledge resources serve as a resource moderating the knowledge conversion. Consequently, the targeted method for the analysis of knowledge characteristics should cover all of the *three* aforementioned *roles of knowledge* for a given business activity.

Requirement 6.4: Equally, the characteristics of the *involved types of agents* (human vs. automated, individual vs. collective), and the *media primarily used for the interactions* (face-to-face vs. virtual) will have to be considered.

Requirement 6.5: The dynamics of knowledge is a third aspect to be included by the analysis method. For each business activity, (a) the *specifics of knowledge conversion* (individual: generation of product representation, transformation of product representation, evaluation of product representation, internalization, externalization; organizational: SECI) taking place, and (b) the *knowledge flows between activities* should be covered by the analysis.

Departing from the idea of dedicated perspectives for the modeled process (Presley, Huff, et al. 1993; Whitman, Huff, et al. 1999; Deng 2007), for each of the aforementioned three knowledge traits a matching process modeling view could be selected. The *functional view* for instance, would be the most appropriate one to describe the knowledge conversion activities interconnected by knowledge flows. Moreover, an extended version of the *resource view* should be capable to capture the involved knowledge resources. Finally, a view derived from the *organization view* might describe the organizational contexts of the business activities.

In contrast, Hommes (2004) criticizes the aforementioned conception of process modeling perspectives for the following reasons:

(a) Ideally, the various process-modeling perspectives should provide orthogonal views on the modeled business process. In practice, however, the different categories of views represent their specific systems directly: An *information view*, for instance, captures essentially the information systems involved in the business process.

(b) Moreover, such perspectives may simply reflect the applied modeling paradigm: A *rules perspective*, for instance, may simply follow a business rule paradigm.

For reasons of proper conceptualization, Hommes (2004) introduces (a) basic perspectives (*structural* and *behavioral*) within each of the identified system categories (*business system, information system, documental system,* and *communication system*), and (b) re-labels the conventional views as distinct *aspect models*. In case of *Event Driven Process Chains* (EPCs), for instance, he extracts four aspect models: *process, data* (based on entity-relationship modeling), *function,* and *organization*.

Requirement 6.6: The analysis of knowledge characteristics primarily acts on the *knowledge system*, which consequently needs to be added to the system categories identified by Hommes (2004). Departing from the notation of aspect models, for each of the requirements 4.2-4.4 a *distinct aspect model* for each of the following characteristics should be included within the analysis method:

    (a) Knowledge resources
    (b) Types of agents and media used for the interactions
    (c) Knowledge dynamics

Requirement 6.7: The analysis method should provide a clear *methodology* describing the required steps and their results for establishing the analysis model on the knowledge level for a selected product development process.

Requirement 6.8: Best practice in process modeling recommends decomposing large models consisting of more than 50 elements in smaller ones (Mendling, Reijers, et al. 2010). This measure helps to (a) improve the ability to understand the process model as a whole, and to (b) reduce errors arising from an error probability higher than 50 % for models comprising more than 50 elements (Mendling, Neumann, et al. 2007). Likewise, the analysis method should offer means for an *adaptable decomposition* at the level of the analyzed process elements and the associated knowledge flows.

6.4 Approaches to the Analysis of Knowledge Characteristics within Product Development

In the year 2000, Nonaka et al. (2000) stated a lack of suitable methods for the evaluation of knowledge resources, although a variety of approaches had been proposed by then. By means of a literature search, focusing on the period since then, relevant analysis and modeling approaches of knowledge characteristics applicable to the product development process will be identified. Subsequently, the uncovered approaches will be evaluated and compared based on the requirements collected in the previous section.

The literature search concentrates on the domains of knowledge management and knowledge science and leaves out (a) approaches primarily targeting the economic assessment of knowledge resources as for instance intellectual capital reporting (Alwert 2005), (b) methods for measuring the return on knowledge (Housel and Bell 2001), or (c) method providing guidance for the knowledge engineering process (Neubert 1993; Schreiber 2000; Stokes 2001).

6.4.1 Knowledge Audit

The *knowledge audit* typically serves as the starting point for a knowledge management initiative in an organization. During this initial step, the knowledge audit will determine (a) which knowledge resources are required for the considered business processes, (b) which gaps need to be closed between the required and the existing knowledge resources, (c) where the knowledge originates from, and (d) who applies which knowledge resources (Liebowitz, Rubenstein-Montano, et al. 2000). The proposed scoping and approaches for knowledge audits differ largely between different authors (Gourova, Antonova, et al. 2009) and therefore a lack of consensus on a widely accepted knowledge audit reference process has to be stated. As Gourova et al. (2009) already present a broad overview on the many proposed approaches for a knowledge audit, the present thesis will only provide a brief introduction to selected approaches.

According to Choy et al. (2004), the complete knowledge audit process comprises three main stages, whereby the first stage prepares the audit by presenting the overall approach to the participating organizations and by a preliminary assessment of the current organizational culture with regards to the readiness for knowledge management. During the second stage, interviews with selected people of the considered business units are conducted by means of surveys. Finally, the interview results are further analyzed in order to obtain the following results during the third stage (Choy, Lee, et al. 2004):

(a) *Knowledge inventory*: Consists of the knowledge resources required by specific roles carrying out a particular business process.

(b) *Knowledge map*: Visualizes the relationships between the business activities, the exchanged knowledge, and the involved individuals.

(c) *Knowledge flow analysis*: Here, social network analysis is applied to determine the critical knowledge providers and consumers.

Nissen (2006) compiled the most relevant steps within the latter two phases of the knowledge audit in the following table:

Table 6.4:   Steps of a knowledge audit, according to (Nissen 2006)

Steps
1. Determine existing and potential knowledge sinks, sources, flows, and constraints.
2. Identify and locate explicit and tacit knowledge.
3. Build a map of the stocks and flows of organizational knowledge.
4. Perform a gap analysis to determine what knowledge is missing.
5. Determine who needs the missing knowledge.
6. Provide recommendations to management regarding necessary improvements.

Apparently, the targeted level of detail within the concluding analysis phase differs largely between the two presented approaches. Nevertheless, the two approaches conform concerning the overall approach and the high-level objectives.

6.4.2 Knowledge Mapping

The application of knowledge maps is not limited to the knowledge audit: In a broader context, knowledge maps may serve as visual directories for all kind of knowledge-related resources in an organization. Here, the overall goal is to provide a visual architecture of organizational knowledge (Eppler 2003) that may be consulted in an interactive way. By extending the ideas of yellow pages that constitute a directory for experts and their set of skills, knowledge maps contain graphical information on the following aspects in an easy to access and navigate way (Eppler 2003):

(a) *Knowledge sources*: experts and their skills
(b) *Knowledge resources*: skills by expert and organizational units
(c) *Knowledge structures*: global architecture of a knowledge domain
(d) *Knowledge applications*: types of knowledge required for a particular business activity
(e) *Knowledge development stages*: stages for the development of a particular competence for individuals or groups

Accordingly, Eppler (2003) proposes five types of knowledge maps where each depicts one of the aforementioned aspects of knowledge. As there is no obligation to use a dedicated modeling language or a specific visualization technique for these maps, knowledge maps feature a rather low level of formalization (Schauer and Schauer 2008).

6.4.3 Knowledge Flow Analysis and Modeling (KFAM)

Newman et al. (2010) propose an analysis and modeling method targeted at the identification of knowledge flows in close association with the business actions and decisions enabled by them. The base ontology of this approach consists of the three concepts (a) *agents*, (b) *knowledge artifacts*, and (c) *knowledge transformations*. The relationships in the ontology originate from the two assumptions that (a) the involved agents perform transformations on artifacts, and that (b) knowledge flows consist of a set of knowledge transformation activities (Newman 2003).

The model foresees three categories of participating agents: *individual agents*, *collective agents*, and *automated agents* (Newman 2003). In a broad sense, the term *knowledge artifact* includes both the physical representation of knowledge (objective) as well as the internalized, cognitive side of knowledge (subjective) (cf. section 3.1). Newman et al. (2010) classify the various types of knowledge-related transformations (e.g. access, search, systematize, discover, codify, store) into the four basic categories comprising (a) *knowledge acquisition*, (b) *knowledge utilization*, (c) *knowledge transfer*, and (d) *knowledge retention*. Here, knowledge utilization plays a singular role as it directly associates with the activities of the business process. Therefore, the knowledge utilization activity serves as a binding element between knowledge flow and business process model. From the business process perspective, the remaining three categories of transformations only play a supporting role.

The KFAM model extends the well-known IDEF0[32] business process model by entities describing the involved knowledge resources and agents. As shown in Figure 6.11, the actual knowledge transformation is represented as activity in the IDEF0-style.

---

[32] IDEF0 is a functional modeling language applied to the analysis and modeling of processes in systems and organizations. It was derived from the *Structured Analysis and Design Technique* (SADT) methodology.

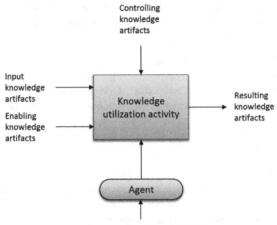

Figure 6.11: Knowledge utilization process element within KFAM, according to (Newman, Conrad, et al. 2010)

The knowledge transformation activity converts the *input knowledge* into the *resulting knowledge*, whereas the *enabling knowledge* serves as a resource allowing this transformation. As an equivalent to controls in the IDEF0 model, the *controlling knowledge artifacts* restrict the transformation activity by providing procedures and policies. Lastly, a dedicated shape directly associated with the transformation activity depicts the agent performing the knowledge transformation. The agent's experiences, skills, and abilities serve as resources for the transformation activity.

Overall, the KFAM method covers the major aspects of the descriptive model of knowledge creation in interdisciplinary product development introduced in section 6.1.3.

6.4.4 Process-based Approach to Knowledge-Flow Analysis

Kim et al. (2003) propose a six-stage procedure for the process-based analysis of knowledge flows that consists of the following steps:

1. Definition of *knowledge ontology* relevant to the analyzed business process
2. *Process analysis*
3. *Knowledge extraction*
4. *Knowledge-flow analysis*
5. *Knowledge specification*
6. *Knowledge validation*

During the *process analysis* step, the business process is (a) decomposed into the relevant activities that are subsequently structured according to their hierarchy and visualized in a process-hierarchy diagram. In the following, (b) the dependencies of the identified business activities are analyzed and depicted in a process-dependency diagram. For each of the identified business activities, three types of knowledge are extracted:

(a) Prerequisite knowledge
(b) Applied knowledge during the execution of the activity
(c) Resulting knowledge

Based on the results of the two preceding steps, *knowledge-flow analysis* establishes a knowledge-flow model consisting of the process activities, the connecting knowledge flows, four types of junctions (join, division, replication, recursion), and the types of knowledge involved. During the subsequent knowledge specification step, the mandatory and optional attributes of the involved knowledge are described. Finally, the obtained modeling items are validated through review sessions and walk-throughs.

As a limitation, the proposed approach does not cover the further decomposition of knowledge flows.

6.4.5 Knowledge Modeling Description Language (KMDL)

Gronau (2012) adopts the model of organizational knowledge creation and *ba* (Nonaka, Toyama, et al. 2000) as a theoretical basis for his modeling framework of knowledge-intensive business processes. At the core of the modeling framework, he introduces the *knowledge modeling description language* (KMDL) that belongs to the family of semi-formal modeling languages. Pogorzelska (2009) emphasizes the better suitability of semi-formal languages when the communication of models between modeling experts and domain experts is required. The semantics of a semi-formal language, however, is not fully defined and does not enforce unambiguousness.

As of version 2.2, KDML focuses on the modeling of various types of conversions between knowledge objects that take place in the context of a business process (Pogorzelska 2009). In the first step, the knowledge-intensive business process is modeled in the *process view* based on an aspect model for the tasks and roles comprising the business process. Using these entities, the process view captures the process flow and the organizational relationships (Pogorzelska 2009). In the second step, the *activity view* enables the modeling of the associated knowledge conversions and flows required for the execution of the various tasks of the business process. Task objects in the activity view that mark several elements as belonging to a specific business task support the mapping between the two views. This view comprises the information and knowledge artifacts involved, individuals and teams, and the actual knowledge conversion flows. Finally, the *communication view* captures the flow of communication between individuals and teams, the applied communication media, and the involved information systems.

Figure 6.12 depicts the knowledge and information objects as well as the occurring knowledge flows for an externalization activity (cf. SECI model in section 5.1.2) in an activity view. On the left, the two knowledge objects (tacit) serve as inputs of the externalization activity that uses the *"Documenting"* technique for the transformation of tacit knowledge into explicit forms. The information object, shown on the right, stores the externalized knowledge within an information object.

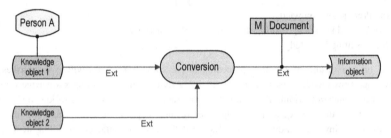

Figure 6.12: Externalization conversion depicted in KMDL's activity view, according to (Pogorzelska 2009)

Although the overall modeling framework clearly distinguishes tacit and explicit knowledge, within KMDL a *knowledge object* represents tacit knowledge only, whereas *information objects* store explicit knowledge (Pogorzelska 2009). Because of this conceptualization, artifacts of explicit knowledge cannot be clearly distinguished from common information objects considered in information flows. Overall, the question whether explicit knowledge may be distinguished from information has been controversially discussed in the research literature. Nissen (2006) proposes an unambiguous distinction of explicit knowledge and information based on the *actionability* or *applicability*. Alternatively, van Renssen (2005) introduces a distinction of information and knowledge based on the content of the considered facts. He labels facts about *individual things* (e.g. instances, objects, components) as information, whereas facts about *kind of things* (e.g. concepts, classes, products) are classified as knowledge.

KDML enables a very detailed modeling of knowledge conversion activities involving the following syntactical elements:

(a) The input and output information and knowledge objects involved in the conversion
(b) The incoming and outgoing edges represent specific knowledge conversions according to the SECI-model. In addition, the specific knowledge conversion method (e.g. practicing, communication, documenting) may be indicated for an edge.
(c) The node representing the knowledge conversion activity is merely the focal point of the incoming and outgoing edges representing the actual conversions.

This way, KMDL permits the modeling of (a) four *atomic* conversions (SECI) having exactly one incoming and outgoing edge of the same type, (b) *complex* conversions having either m:1 or 1:n input and output objects, and (c) *abstract* conversions having several (m:n) input and output objects.

Overall, the KMDL modeling framework represents the concepts of the model of organizational knowledge creation and *ba* (Nonaka, Toyama, et al. 2000) very closely. One shortcoming arises from the missing ability to handle knowledge resources serving as a resource moderating the knowledge conversion activities. Unfortunately, the KMDL method does not offer sufficient means for an adaptable decomposition. As the modeling in the activity view requires a significant effort, it should only be used in case of knowledge-intensive tasks. In order to support the activities of individual knowledge creation, KMDL has to be extended by (a) dedicated activities and (b) individuals as actors.

6.4.6 State-Process-Resource Modeling

The state-process-resource model departs from the idea that the product development process can be sufficiently modeled based on the following three concepts:

(a) *Product state*: Each product state describes the results constituting a product at the end of a development phase within the overall product development process.

(b) *Process element*: activity in the product development process that transforms a product from $state_n$ to the $state_{n+1}$

(c) *Resource*: They subsume everything needed to execute a process element, e.g. methods, tools, and tacit knowledge applied within the product development process.

In order to properly locate the required resources for the analyzed development process, the product states, process elements, and applied resources have to be identified in the first step, while the second step maps and structures knowledge on these objects (Deng 2007). Consequently, Deng (2007) classifies the knowledge in product development by means of a two-layer model described in section 6.2. The modeling approach includes features for the composition and decomposition of product states based on IDEF5 junction symbols.

The state-process-resource method provides means for an adaptable decomposition at the level of the modeled process elements. It covers, however, only certain aspects of the descriptive model of knowledge creation in interdisciplinary product development introduced in section 6.1.3. While resources can be considered as the representative for knowledge resources in the sense of the descriptive model of knowledge creation, the state-process-resource model does not take into account the involved actors and knowledge flows. Likewise, the process elements do not explicitly describe the underlying knowledge conversion activities.

6.4.7 Assessment of Requirements Coverage

In the following, the requirements established in section 1.1 will be applied as assessment criteria to the previously presented analysis approaches of knowledge characteristics. For the assessment of the requirements coverage, the following symbols will be used:

O   =   no coverage

◑   =   partial coverage

●   =   full coverage

Among the seven analyzed methods, the KFAM and KMDL approaches were evaluated as the best approaches. Both approaches, however, still bear potential to gain a better coverage of the described requirements. Here, KFAM can be further improved by adding (a) distinct aspects model describing knowledge resources, types of agents and knowledge dynamics, and (b) means to model the characteristics of the *ba*. KMDL may benefit from extensions toward the (a) handling of knowledge resources serving as a resource moderating the knowledge conversion activities and (b) means for an adaptable decomposition.

Table 6.5:   Assessment of requirements coverage for analysis methods of knowledge characteristics

Approaches						
ID / **Requirements**	Knowledge Audit	Knowledge Mapping	KFAM	Process-Based Approach to Knowledge-Flow Analysis	KMDL	State-Process-Resource Modeling
6.1 Usage of the activities of the associated product development process as *integration* and *anchor points*	◐	○	●	●	●	●
6.2 Modeling of *knowledge flows not aligned* with the activities of the associated product development process	◐	○	●	◐	●	○
6.3 Modeling of *three roles of knowledge* (input, output, resource for knowledge conversion) for a given business activity	◐	○	●	●	◐	○
6.4 Modeling of the characteristics of the *involved types of agents* (human vs. automated, individual vs. collective), and the *media primarily used for the interactions* (face-to-face vs. virtual)	○	○	◐	○	●	○
6.5 Modeling of the *specifics of knowledge conversion* (individual: creation of product representation, transformation of product representation, evaluation of product representation, internalization, externalization; organizational: SECI) taking place, and the *knowledge flows between activities*	◐	○	●	◐	◐	○
6.6 *Distinct aspect models* for knowledge resources, types of agents and media used for the interactions, and knowledge dynamics	◐	◐	○	◐	◐	○
6.7 Usage of clear *methodology* describing the required steps and their results for establishing the analysis of knowledge characteristics	●	◐	●	●	●	●
6.8 Support for an *adaptable decomposition* at the level of the analyzed process elements and the associated knowledge flows	○	○	●	○	◐	●

6.5 Synthesis of Research Framework

In the following, the components of the research framework will be described that provide the answer to the second research questions formulated in section 1.2:

> *How might knowledge constituents relevant within product development (knowledge resources, structure of knowledge and knowledge flows) be analyzed? What could be the appearance of a research framework that provides the methodological basis for the analysis of the knowledge characteristics of product development?*

Previously, the KFAM and KMDL methods were selected as the most capable analysis approaches of knowledge characteristics described in the research literature. Both approaches, however, do not suffice to cover the full range of requirements documented in section 6.3. They may therefore only serve as the basis for an extended approach that meets the specified needs. Overall, the KMDL method offers the better basis for the required extensions due to (a) its support for extensibility through concepts like aspect models, and (b) the higher expressiveness of the modeling language. Therefore, KMDL will be chosen as the foundation permitting the following extensions:

(a) Handling of knowledge resources serving as a resource moderating the knowledge conversion activities in the *activity view*

(b) Means for a further dissection of knowledge conversion activities in the *activity view*

(c) Support for individual knowledge creation

KMDL already supports the modeling of various types of requirements directly associated with a conversion activity. Here, requirements represent either (a) needed knowledge of individuals and teams or (b) specific information (Pogorzelska 2009).

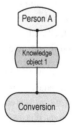

Figure 6.13: Extension of KMDL for knowledge resources moderating the knowledge conversion activity

In a similar way, knowledge resources that moderate the knowledge conversion activity may be directly associated with the conversion activity as shown in Figure 6.13.

In the process view, KMDL already permits the decomposition of the overall business process into a number of sub-processes by means of the *process interface* element that serves as connector tying together the various sub-processes. By default, the activity view models the knowledge conversion activities belonging to the content of the corresponding process view. The number of knowledge conversion activities occurring for one task of the process view, however, may be quite large. Therefore, an *activity interface* element will be introduced that allows for a decomposition of activity views into smaller blocks without artificially increasing the resolution of the process view. Figure 6.14 depicts the activity interface element *"Conversion$_{n-1}$"* as a connector to the preceding knowledge conversion activity, whereas the activity interface element *"Conversion$_{n+1}$"* stands for a connection to the subsequent activity.

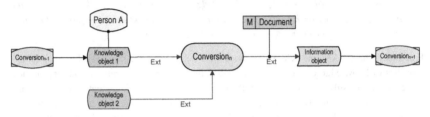

Figure 6.14: Extension of KMDL by activity interfaces

In order to support the activities of individual knowledge creation (cf. section 6.1.1), KMDL will be extended by new conversion activities for the cognitive activities of generation, transformation and evaluation of product representations, whereas for the internalization and externalization conversions at the individual level the existing elements of KMDL will be used. Finally, the difference between the individual and social variants can be grasped from the different types of actors involved (individual vs. group).

As discussed in section 6.4.5, within KMDL a *knowledge object* represents tacit knowledge only, whereas *information objects* store explicit knowledge (Pogorzelska 2009). This modeling approach, however, does not permit to distinguish explicit knowledge in the form of mental representations in the human brain (subjective domain) from articulated, explicit knowledge in its physical form (objective domain). Moreover, the view of the connection between tacit and explicit knowledge as a continuum leads to a differing modeling pattern that employs *knowledge objects* to represent tacit and explicit knowledge in its mental representation, whereas *information objects* are used for articulated, explicit knowledge.

As the final constituent of the research framework, the methodology describing the required steps and their results for establishing the analysis model of knowledge characteristics for a given product development process will be presented. The KMDL method itself provides a large procedure model defining the various phases typically conducted in a KMDL-based consulting and analysis project (Pogorzelska 2009). The overall goal of such projects consists of a very detailed process analysis and subsequent process improvement and evaluation steps. It therefore exceeds the needs for the discussed analysis of knowledge characteristics of a given product development process substantially. Consequently, of the nine phases proposed by KMDL, only three will be used in the research framework under conception as depicted in Figure 6.15.

In the first step, the process model will be established by means of the process view. In a subsequent step, the process model will be examined in order to identify knowledge-intensive tasks. The identified knowledge-intensive tasks will then be modeled in the activity views forming the activity model.

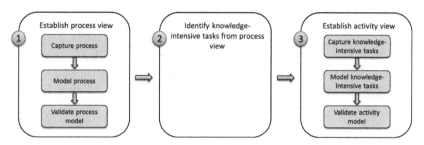

Figure 6.15: Methodology for analysis of knowledge characteristics, adapted from (Pogorzelska 2009)

7 Process Elements of Mechatronic Product Development

In this chapter, the previously established research framework will be applied for the analysis of knowledge characteristics of MPD. For the knowledge-intensive development activities conducted in the context of MPD, the analysis will be able to identify:

(a) Knowledge resources in three distinct roles as input, output and as a resource moderating the knowledge conversion activities
(b) Knowledge contexts including the types of involved agents (human vs. automated, individual vs. collective), the roles of the involved agents and their disciplinary affiliations, and the media primarily used for the interactions (face-to-face vs. virtual)
(c) Knowledge flows and the specifics of the knowledge conversion taking place

For the intended design support system, this analysis will help to capture the available information and knowledge resources embedded in models. At runtime, the system will utilize them to supply its users with the information and knowledge objects required to conduct a particular development activity. Overall, the design support system needs insights on (a) the different working steps conducted within a process element, (b) the information and knowledge objects utilized within these working steps, and (c) the available information and knowledge objects from existing models.

Freisleben (2001) identified a range of process elements common to product development processes independent of the specifics of the company and the involved engineering disciplines. Therefore, it can be safely assumed that this holds also true in the context of the Mechatronic product development processes, where for instance the VDI guideline 2206 (2004) proposes a range of process elements. Consequently, the determined characteristics of the process elements for MPD can be included directly within the design support system as a ready-to-use library of common process elements as e.g. provided within the software proNavigate (Vajna 2006). Nevertheless, the design support system should also be able to handle process elements specific to the particular development process.

Section 7.1 develops the approach for establishing a process view of MPD based on common process elements[33]. Following this approach, section 7.2 establishes the KMDL process view for a selection of characteristic process elements of MPD. Finally, section 7.3 captures the knowledge characteristics of these process elements by means of the KMDL activity view.

In summary, the present chapter will provide the answers to the third and fourth research questions formulated in section 1.2:

3. *Which process elements are commonly applied within MPD?*
4. *What are the knowledge needs of the different disciplines and roles within each of these process elements?*

---

[33] Typically, these process elements consist of multiple working steps that belong to a fixed workflow. Therefore, they are compliant to the definition of a process element proposed by (Freisleben 2001).

7.1 Approach for Establishing a Process View based on Common Process Elements

Within the first step of the three-stage methodology introduced in section 6.5, the different activities conducted in a business process will be captured and modeled by means of the KMDL process view. For this purpose, a comprehensive description of the MPD process has to be established. In section 2.4.2.1, the procedure model of the VDI guideline 2206 (VDI 2004) was characterized as the most extensive procedure model for MPD (Felgen 2007). Therefore the question arises as to whether this procedure model suffices for establishing a process description with (a) a sufficient level of detail and (b) an adequate coverage of the broad spectrum of mechatronic systems. The rather generic nature of this procedure model, however, impedes its direct application as reference process (cf. section 2.4.2.1.5). For this reason, (a) additional process elements and (b) an increased level of detail for already included process elements in this guideline will have to be developed.

This statement has been confirmed by recent research results of Torry-Smith and Mortensen (2011), who analyzed the research literature on the mechatronic design process regarding the coverage of seven highly relevant aspects in MPD:

(a) Synchronization between the Mechatronic process model and the process models of the involved disciplines
(b) Each discipline should be perceived as constantly evolving their disciplinary design in between "integration meetings". Therefore, the parts of the disciplinary design likely to change have to be clarified upfront.
(c) Multiple allocations of functions to disciplines should be investigated in order to determine the "best fit" solution.
(d) Distribution of functions and properties between disciplines
(e) Shared models capturing the allocation of functions and properties to product components
(f) Specification of interfaces
(g) Inclusion of the user-perceived value of the product into the design process

As it turns out, the considered procedure models provide excellent support only for two out of seven considered aspects in mechatronic synthesis that are closely associated with control engineering and partitioning. For other scenarios of MPD, the authors concluded a lack of methodological support. In particular, the analysis rated the procedure model of the VDI guideline 2206 as providing only partial support in just four of seven aspects, whereas no excellent support at all could be acknowledged. These days, even some of the authors of the VDI guideline 2206 consider the proposed procedure model only *"as first step on the way to a holistic design methodology for mechatronic systems"* (Gausemeier and Kahl 2010).

From the analyses described, the following conclusions can be drawn:

(a) Overall, the considered procedure models for MPD do not provide a sufficient coverage of all relevant aspects within the synthesis of mechatronic products.
(b) So far, no comprehensive reference model for MPD could be uncovered that comprises the required process elements for MPD and their sequencing. A reference

model usually contains a superset of process elements and may be used as a blueprint requiring certain adaptations for a concrete process.

Due to the aforementioned arguments, the present thesis will conduct the analysis of knowledge characteristics at the level of process elements that have to be identified, collected, and compiled in a first step. For this purpose, Freisleben (2001) introduced an approach for (a) the identification of process elements from the research literature and (b) the subsequent alignment of the various process elements with regards to scope and level of detail. In order to assure a proper chaining, the process elements are further enriched with a description for the incoming and outgoing information.

In conclusion, Table 7.1 summarizes the chosen approach for establishing the process view of MPD based on common process elements:

Table 7.1:   Approach for establishing a process view for MPD based on common process elements

Steps	Results
1. Identify relevant procedure models of MPD from the research literature	List of procedure models of MPD
2. Identify the relevant process elements from each procedure model	List of process elements for each procedure model
3. Match and harmonize process elements with similar content	Multiple sets of matching and harmonized process elements from all procedure models
4. Define a process element matching each set of similar process elements	List of compiled process elements, where each matches a set of similar process elements of the various procedure models
5. Model the compiled process elements by means of the KMDL process view	KMDL process views for each of the compiled process elements

The described approach offers the advantage of taking into account a wide range of proposed procedure models and their process elements. In addition, it is able to cover the variety of activities performed in the contexts of the various process elements as well as the roles involved in these activities.

7.2 Identification and Compilation of Common Process Elements for MPD

As described by the first step of the three-stage methodology introduced in section 6.5, the process view of MPD will be established based on common process elements. For this purpose, the five steps of the approach conceived in the previous section will be conducted. Using the results of this process, this section will provide the answer to the third research question formulated in section 1.2:

3.   Which process elements are commonly applied within MPD?

7.2.1 Identification of Relevant Procedure Models of MPD

In the first step, relevant procedure models described in the research literature will be identified:

- Bellalouna (2006) describes an integration platform for the interdisciplinary development of mechatronic products based on a service-oriented architecture (SOA). The software platform focuses on the realization of the interdisciplinary uses cases within MPD. Deviating from the procedure model described in the VDI guideline 2206, Bellalouna (2009) follows the specific methodology of automotive engineering and consequently adopts the logical and technical system architecture as described in section 2.6.3. In addition, he takes into account the versioning processes for components, domain-specific systems, and at the level of the overall system.
- Follmer, Hehenberger, et al. (2011) propose a procedure model for model-based mechatronic design that uses the V-model of VDI guideline 2206 as base for integration. Overall, the approach (a) largely replaces the design steps of the V-model by five phases adopted from the VDI guideline 2221 and (b) describes approaches for establishing a mechatronic system model (MSM) in bottom-up and top-down modeling.
- Gausemeier and Kahl (2010) introduce a new design methodology for mechatronic and self-optimizing systems. It proposes a procedure model built from the two main phases of (a) conceptual design and (b) concretization.
- The ISYPROM Consortium (2011b) proposes a procedure model for system development that builds up on various procedure models from product development (VDI 2206, VDI 2221) and systems engineering. The described procedure model covers the complete product lifecycle from planning, over design, realization, manufacturing, to utilization.
- Jansen (2006) elaborates on the activities of functional and spatial partitioning and describes a partitioning methodology complemented by process elements, strategies and partitioning rules.
- For a detailed description of the procedure model proposed by VDI 2206 (2004) see section 2.4.2.1.

Other relevant procedure models were taken into consideration but could not be included in the identification of process elements due to the following reasons:

- Hellenbrand and Lindemann (2011) describe a framework for the process modeling and planning of mechatronic products, where the functional validation of the overall system function during the phases of integration and testing form the basis for the planning of the development process. The applicability of the framework is limited to cases where the overall system elements are already known and can therefore not be applied to the development of new products.
- The applicability of the W Model proposed by (Nattermann and Anderl 2011) is limited to adaptronic products.

7.2.2 Clarification of Terminology and Alignment of Process Elements

Overall, the three process modules of the VDI guideline 2206 were used as a structuring frame with the addition of a new category of recurring activities encountered in multiple process modules. The analysis of the process elements proposed by the various considered procedure models revealed large differences in (a) the applied terminology and (b) the proposed process elements.

Gausemeier and Kahl (2010) use the term *"principle solution"* to describe the result of the system design phase comprising the overall structure and the modes of operation of the mechatronic system. Subsequently, the principle solution will be employed for further concretization in the domain-specific design phase. In their opinion, the principle solution comprises among others the aspects of requirements, functions, Wirk-structure, shapes, and behavior. The VDI guideline 2206 favors the term *"cross-domain solution concept"* instead. Follmer, Hehenberger, et al. (2011) introduce the term *"mechatronic system model"* referring to a representation of the mechatronic system that comprises its main properties, component structure, requirements, functions, behavior, and existing CAx-models. They consider the principle solution only as an intermediate result of the principle design stage. In contrast, the procedure model proposed by the ISYPROM Consortium (2011b) favors the designations (a) *"preliminary design"* comprising the function structure, the break-down of the system in modules, solution elements and Wirk-structures for the various modules, and (b) *"system design"* referring to a detailed definition of the system consisting of function structure, Wirk-structure, component structure and interface definitions for the various modules.

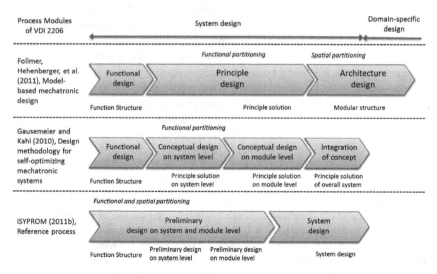

Figure 7.1: Comparison of proposed procedure models within the system design stage

Figure 7.1 depicts the proposed sub-processes as well as their results by three of the most elaborate procedure models for the system design phase. The diagram clearly underlines that the employed wording for the different sub-processes and their results differ significantly.

For the identification of common process elements within these sub-processes, these terminological differences have to be streamlined. Therefore, several commonly used designations for process elements and their results will be introduced as indicated in Table 7.2. Appropriately, the proposed terminology will be used in the common process elements compiled in the table of appendix 1.

Table 7.2:   Common process elements and their associated results within the system design stage

Process elements	Results
Establish function structure	Function structure
Conceptual design at system level (functional partitioning)	System design concept: - Function structure - Wirk-principles/solution elements - Wirk-structure
Spatial partioning into subsystems and components resulting in component structure	Preliminary component structure
Conceptual design at subsystem level	Subsystem design concept: - Function structure of subsystem - Wirk-principles/solution elements of subsystem - Wirk-structure of subsystem
Concept integration and interface definition	System design: - Function structure - Wirk-principles/solution elements - Wirk-structure - Component structure - Definition of interfaces

7.2.3 Compilation of Common Process Elements and Roles

The intended process description may vary in the number of considered disciplines. For instance, when modeling the development process from the perspective of integrated product development (Andreasen and Hein 1987; Ehrlenspiel 1995), activities of departments such as marketing and sales have to be included into the process model. Consequently, the resulting process model will grow significantly in size and complexity. As a result, the subsequent modeling of knowledge-intensive activities in the activity view would exceed the scope of the present research due to the laborious nature of the modeling of the involved knowledge conversions. Therefore, the present thesis will focus on (a) the core product development activities only (i.e. the scope described by three process modules described of VDI guideline 2206) and (b) the previously identified constituents of mechatronics, i.e. mechanical engineering, electrical engineering, and computer science with control engineering as one of its sub-disciplines (cf. section 2.1.2).

Before elaborating on the results obtained from the steps 2-4 listed in Table 7.1, a couple of assumptions used within this process will be laid out:

(a) Following the approach used in the VDI guideline 2206, previously established requirements form the input of the system design process module. Accordingly, the process elements of task clarification and requirements design precedent to system design will not be considered.

(b) The specifics of the approach proposed by Gausemeier and Kahl (2010) regarding the handling of multiple specific application scenarios were not considered, as they significantly differ from the other approaches.

(c) In order to avoid reaching a level of complexity difficult to handle, only the case of implicit partitioning will be considered.

(d) The fourth step of the approach laid out in Table 7.1 results in the common process elements extracted from the various process elements proposed by the considered procedure models listed in the table in appendix 1. This table uses the three process modules of the VDI guideline 2206 and an additional category of recurring activities encountered in multiple process modules as its overall structure.

The results of a similar analysis of the considered procedure models on the roles proposed for MPD are compiled in appendix 2. However, a proper understanding of certain roles mentioned in this table may require additional explanations (ISYPROM Consortium 2011b):

- *System architect*: Establishes the high-level architecture of the overall system, partitions the system in subsystems and takes care of the analysis and assessment of variants of the conceptual design on the system and subsystem level.

- *Product/system manager*: Acts as the owner of the product over all phases of its lifecycle.

- *Head of function*: Takes care for the detailing of specific functions within a particular domain.

- *System analyst*: During the system design phase, he/she assures that the overall system and the comprised subsystems meet the specified requirements. Furthermore, he/she establishes an analysis model, the use cases of the system and finally conducts the analyses of the system.

7.2.4 Establishing the KMDL Process View for Common Process Elements

As the last of the five process steps listed in Table 7.1, the previously compiled process elements will be modeled by means of the KMDL process view. Ideally, the process view should be established for all of the common process elements of MPD. Due to the space required to present and document the obtained results, the process modeling activities will be confined to a selection of process elements being representative for the overall set of process elements.

This selection of typical process elements should comprise items covering the following aspects:

(a) Mono-disciplinary and interdisciplinary development activities
(b) Individual and organizational knowledge creation activities

Moreover, the subsequent modeling of the KMDL activity view for these process elements is useful for knowledge-intensive tasks only. Therefore, it seems reasonable to consider only knowledge-intensive items in the selection of typical process elements. For this purpose, the selection process will incorporate the identification of the knowledge-intensive tasks as defined by the second step of the three-stage methodology introduced in section 6.5. Within this step, the characteristics of knowledge-intensive business processes described in the introduction of chapter 6 will serve as criteria for the identification of knowledge intensive tasks. Table 7.3 depicts the characterization of process elements according to these criteria obtained from the identification process of knowledge-intensive process elements.

Table 7.3:   Identification of knowledge-intensive process elements

Process elements	Criteria			
	High contribution of knowledge to the process' added value	Large share of creative parts within the process	Emphasis on communication	Build-up of new knowledge is time and resource intensive
Derive functions from requirements	●	●		
Establish function structure	●	●		
Conceptual design at system level (functional partitioning)	●	●	●	●
Spatial partioning into subsystems and components resulting in component structure	●	●	●	●
Conceptual design at subsystem level	●	●		●
Analyze and assess variants for conceptual design of subsystems	●	●		●
Establish and verify one or multiple variants of the system design concept	●	●	●	●
Select one of the variants of the system design concept	●	●	●	
Derive domain-spanning system design	●	●	●	●
Derive domain-specific function structures	●			
Establish E/E system architecture	●	●	●	●

Establish software system architecture	•	•	•	•
Establish mechanical system architecture	•	•	•	•
Stepwise integration of components and subsystems to overall system	•		•	•
Conduct integration tests of subsystems and overall system	•			•
Identify incompatibilities between subsystems	•			•
Eliminate incompatibilities between subsystems	•		•	•
Determine optimal overall solution according to requirements	•	•	•	•
Establish software baseline			•	
Establish E/E baseline			•	
Establish mechanical configuration			•	
Establish configuration of integrated system			•	
Modeling and model analysis	•	•	•	•
Assurance of properties according to requirements	•		•	•
Version interdisciplinary system			•	
Release interdisciplinary system			•	

As previously explained, the subsequent selection of typical process elements will consider only such process elements, for which two or more of the criteria in Table 7.3 apply. For those knowledge-intensive process elements, Table 7.4 depicts the results of the subsequent characterization of typical process elements.

Here, the classification according to the first criteria was established by roughly answering the question, which of the roles identified in section 7.2.3 appear in the given process elements. The exclusive appearance of disciplinary component developer roles, as e.g. mechanical design engineer, E/E design engineer, and software developer, in a process element, indicates mono-disciplinary development activities. In contrast, interdisciplinary process elements can be identified from (a) the appearance of roles adhering to different disciplines and/or (b) the involvement of system-level roles as e.g. system architect, system analyst, and system integrator. Table 7.4 shows the resulting distribution of mono-disciplinary and interdisciplinary process elements, where the system design and system integration phases comprise only interdisciplinary process elements, whereas the disciplinary design phase by definition consists solely of mono-disciplinary activities. A generally valid disciplinary characterization of the two activities encountered in multiple process modules (*Modeling and model analysis, Assurance of properties*) is not feasible, as their disciplinary focus directly depends on the types of models being considered and the disciplinary adherence of the properties to be assured.

Table 6.3 described the traits of individual and organizational knowledge creation in product development in detail. These characteristics will be adopted for the classification of knowledge-intensive process elements according to the second criteria that captures the knowledge conversion mode. For this classification, the following cases can be distinguished:

(a) If either an individual or multiple, non-interacting individuals conduct a process element, only *individual knowledge creation* can take place.

(b) If conducting non-creative tasks, only experiential learning might take place that eventually leads to a built-up of tacit knowledge. In addition, *organizational knowledge creation* might be necessary e.g. to share tacit knowledge between individuals. Table 7.4 does not consider this scenario due to the focus on knowledge-intensive process elements.

(c) If working on creative tasks in a team context, typically a combination of *individual and organizational knowledge creation* will occur (cf. Figure 6.5).

Table 7.4:   Characterization of process elements by two main aspects

Knowledge-intensive process elements	Domain scope		Knowledge conversion mode	
	Mono-disciplinary	Interdisciplinary	Individual knowledge creation	Organizational knowledge creation
Derive functions from requirements		●	● (transformation + evaluation)	●
Establish function structure		●	● (transformation + evaluation)	●
Conceptual design at system level (functional partitioning)		●	● (transformation + evaluation)	●
Spatial partioning into subsystems and components resulting in component structure		●	● (transformation + evaluation)	●
Conceptual design at subsystem level		●	● (transformation + evaluation)	●
Analyze and assess variants for conceptual design of subsystems		●	● (evaluation)	●
Establish and verify one or multiple variants of the system design concept		●	● (transformation + evaluation)	●

Select one of the variants of the system design concept		●	● (evaluation)	●
Derive domain-spanning system design		●	● (transformation + evaluation)	●
Derive domain-specific function structures	●		● (transformation + evaluation)	●
Establish E/E system architecture	●[34]		● (transformation + evaluation)	●
Establish software system architecture	●		● (transformation + evaluation)	●
Establish mechanical system architecture	●		● (transformation + evaluation)	●
Stepwise integration of components and subsystems to overall system		●	●	●
Conduct integration tests of subsystems and overall system		●	● (experimentation + evaluation)	●
Identify incompatibilities between subsystems		●	● (validation)	●
Eliminate incompatibilities between subsystems		●	● (transformation + evaluation)	●
Determine optimal overall solution according to requirements		●	● (evaluation)	●
Modeling and model analysis	n/a		● (transformation + evaluation)	n/a
Assurance of properties according to requirement	n/a		● (evaluation)	n/a

For the process elements of three process modules of VDI guideline 2206, Table 7.4 shows that individual and organizational knowledge creation appear in combination. For the two activities encountered in multiple process modules (*Modeling and model analysis, Assurance of properties*), the occurrence of organizational knowledge creation depends on the types of models being considered and the disciplinary adherence of the properties to be assured.

---

[34] The E/E system architecture comprises electrical and electronics components in an interwoven manner. Establishing the E/E system architecture is typically considered as a mono-disciplinary activity as both electronics and electrics adhere as subdisciplines to electrical engineering.

Based on the achieved classification, the following process elements will be selected as the representative ones for all of the knowledge-intensive process elements covering the three process modules of VDI guideline 2206:

(a) *System design*:
  i.   Derive functions from requirements
  ii. Establish function structure
  iii. Conceptual design at system level (functional partitioning)

(b) *Disciplinary design*:
  i.   Derive domain-specific function structures from system-level function structure
  ii. Establish software system architecture

(c) *System integration*:
  i.   Stepwise integration of components and subsystems to overall system

Appendix 3 compiles the KMDL process views established for these process elements.

7.3 Knowledge Characteristics of Common Process Elements

By following the final step of the three-stage methodology introduced in section 6.5, the previously selected six process elements will be modeled by means of the KMDL activity view. Based on the characteristics captured within the KMDL activity views, the knowledge specifics of each process element can be extracted and documented. Based on these results, this section will provide the answer to the fourth research question formulated in section 1.2:

4. What are the knowledge needs of the different disciplines and roles within each of these process elements?

7.3.1 Establishing the KMDL Activity View for Common Process Elements

Appendix 4 compiles the KMDL activity views established for the six process elements, where each sheet of the activity view captures the knowledge conversions associated to a specific task of a process element. All considered process elements employ multiple roles: Typically, a specific role conducts the transformation activities (e.g. system architect, software architect, system integrator), whereas the subsequent evaluation activities are performed by another role (e.g. system analyst). Due to the appearance of multiple roles per process element, performing these tasks typically requires a combination of *individual* and *organizational knowledge conversion* activities as stated in the previous section. In the following, the specifics of the modeling of organizational vs. individual knowledge conversion activities will be outlined.

The task *"Search appropriate Wirk-principles/solution elements"* in the context of the process element *"Conceptual design on system level"* may serve as an illustration for the application of organizational knowledge creation. Here, the modeling of the conversion activities employs all four conversion modes described by the SECI-model. The task starts with an internalization conversion, then employs socialization during brainstorming, externalizes the ideas obtained from the brainstorming, and finally combines the externalized ideas with information from design catalogs. This example illustrates that the four modes of knowledge conversions in the organizational context are well suited to model interaction scenarios of team members.

The resulting activity view, however, appears to be rather complex. In order to reduce its complexity level, the internalization conversion might be omitted. An internalization conversion is typically required to embody explicit knowledge originating from information objects into an individual. In general, an internalization conversion may require specific knowledge resources of the involved actor. In the context of the considered process elements, however, the knowledge required for internalization is part of the knowledge profile of the executing roles and would therefore not specifically contribute to the analysis of the knowledge characteristics.

The process element *"Establish software system architecture"* contains several transformation tasks executed solely by the software architect. It may therefore serve as an illustration of the specifics within individual knowledge creation. In the first task *"Identify logical software components for software functions"*, the software architect starts by generating ideas for a logical grouping of the software functions contained in the software function structure. Next, he/she structures the software functions according to the proposed logical grouping and associates them with logical software components. This concretizing activity transforms the software function structure at the input into a more concrete preliminary functional system diagram. In a subsequent addition activity, data objects and data flows are added to the diagram leading to the final functional system diagram. In the context of individual knowledge conversions, the standard control flows of KMDL associated with the conversion modes of the SECI-model cannot be applied. Consequently, the control flows are depicted as black arrows without a label indicating an *"undefined conversion"* according to KMDL.

7.3.2 Conclusions

Altogether, the proposed research framework for the analysis of knowledge characteristics in MPD proved to be capable of modeling all knowledge conversion activities for the six considered process elements. Activities carried out by individuals as well as those by teams could be adequately described. Thereby, the knowledge conversion modes proposed by the integrated descriptive model of knowledge creation in interdisciplinary product development covered all of the analyzed activities. The three types of cognitive activities on representations (generation, transformation, and evaluation) described by (Visser 2006a) were employed to capture the high-level characteristics of each task modeled in the activity view. At the level of a specific activity, however, the types of detailed transformation activities (e.g. addition, detailing, concretizing) and several types of supportive cognitive activities (e.g. analysis, hypothesizing) were used additionally. The syntactic elements of the extended KMDL modeling method were appropriate for the modeling of the considered process elements in the process and activity view.

The activity views established by completing the three-stage methodology described in section 6.5, will be used to extract and document the knowledge characteristics of each process element.

Table 7.5: Knowledge characteristics determined for common process elements

Knowledge-intensive process elements	Knowledge characteristics				
	Used roles	Used resources from IT systems	Input information and knowledge objects	Used knowledge resources	Output information and knowledge objects
Derive functions from requirements	- System architect - System analyst		- Requirements specification		- Descriptions of overall functions, main functions, auxiliary functions - Description of requirements coverage
Establish function structure	- System architect - System analyst		- Descriptions of overall functions, main functions, auxiliary functions		- Function structure - Description of achieved coverage
Conceptual design at system level (functional partitioning)	- System architect - System analyst - Head of function	- Design catalogs	- Function structure	- Understanding of already realized solutions - Knowledge on compatibility of Wirk-principles/solution elements	- System design concept
Derive domain-specific function structures	- System architect - Head of function		- System design concept	- Knowledge on domain adherence of Wirk-principles/solution elements	- Mechanical function structure - E/E function structure - Software function structure
Establish software system architecture	- Software architect - Head of function		- Requirements specification - System design - Software function structure		- Functional system diagram - Component diagram - Deployment diagram - Description of achieved coverage
Stepwise integration of components and subsystems to overall system	- System integrator - System analyst - Head of function		- System design - Mechanical function structure - E/E function structure - Software function structure		- Interdisciplinary product structure

Table 7.5 compiles the knowledge characteristics determined for the analyzed process elements based on the following categories:

(a) Roles appearing in the activity views
(b) Utilized information and knowledge resources from IT systems
(c) Input information and knowledge objects
(d) Knowledge resources moderating the knowledge conversion activities
(e) Output information and knowledge objects

The captured information and knowledge objects depicted in the table do not include background knowledge and knowledge commonly associated with the appearing roles. If required, the level of detail on the included information and knowledge objects could be further increased by refining the analyzed activity views. As a whole, the knowledge characteristics described in Table 7.5 constitute the answer to the fourth research question formulated in section 1.2.

8 Semantic Technologies for Design Support in Mechatronic Product Development

> *I have a dream for the Web [in which computers] become capable of analyzing all the data on the Web – the content, links, and transactions between people and computers. A "Semantic Web", which makes this possible, has yet to emerge, but when it does, the day-to-day mechanisms of trade, bureaucracy and our daily lives will be handled by machines talking to machines.*

> Berners-Lee and Fischetti, 1999

In the above quote, Tim Berners-Lee, the inventor of the *World Wide Web*, lays out his vision for complementing the existing web of human-readable HTML pages by a web of machine-understandable information that can be shared and reused by applications. Similar to the aforementioned HTML pages of the current web, the manifold types of models constitute the main carriers of information in the context of product development. Although models are considered an important means for communication within the design process (Buur and Andreasen 1989), their content is understandable in most cases only to the associated authoring tool. As for the *Semantic Web*[35], the access to the information stored in the various types of models is essentially a *data integration* problem. For this type of problem, semantic technologies can be employed to make the information embedded in models' data understandable to a greater audience. By means of common vocabularies (i.e. metadata models or shared conceptualizations), the interlinked data from these heterogeneous sources can be queried comparable to a large knowledge base providing access to the abundance of information and knowledge on the product provided by the created models.

In addition, the information provided by these models can be presented to users according to their specific knowledge profile, i.e. for specialists in a certain domain, the full range of available information can be presented, whereas users from other domains will be presented only a filtered view of the information understandable to them.

Section 8.1 gathers the high-level requirements for the envisioned design support system. Later on, the gathered requirements will guide the selection of the semantic technologies to be employed and the conception of the architecture of the design support system. Next, section 8.2 gives an overview of the semantic technologies utilized for the *Semantic Web*. Finally, section 8.3 describes the application of semantic technologies for the envisioned design support system in MPD. In summary, the present chapter will answer the fifth research question formulated in section 1.2:

---

[35] The W3C describes the vision of the *Semantic Web* and provides access to the underlying standards under: http://www.w3.org/standards/semanticweb/

5. *What are adequate semantic technologies permitting a semantic enrichment of the numerous types of models used within MPD allowing to improve the processing, storage, distribution and context-sensitive provisioning of knowledge?*

8.1 Requirements for the Design Support System in MPD

This section will collect the requirements for the envisioned design support system. These requirements will subsequently guide both the selection of the semantic technologies to be employed and the architectural design of the design support system. The descriptive model and the associated modeling framework developed in the two preceding chapters provide the conceptual basis for the design support system. In very general terms, the design support system aims at improving the creation, storage, distribution, and context-sensitive provisioning of information and knowledge throughout the widespread and complex networks encountered in MPD. In the following, this high-level idea will be broken down into particular requirements.

Requirement 8.1: The parts of common interest in the native model data should be made understandable to other participants of the product development process who do not possess the authoring tool. For this purpose, the native model data should be *attributed* by *metadata* originating from discipline-specific *vocabularies*, i.e. common metadata models or shared conceptualizations.

Requirement 8.2: While the first attribution step explicitly declares the relationships between the internal resources within a model, in the next step these internal resources should be *linked* to *external resources* based on the semantic vocabulary. Consequently, networks of interlinked information between the models will be established.

Requirement 8.3: The chosen vocabularies should provide access to the information and knowledge embedded in the model to applications sharing the same vocabularies. In addition, these *semantically enabled applications* will also be able to process this information and knowledge.

Requirement 8.4: These networks of interlinked information and knowledge can be exploited to provide each user appropriately with the information and knowledge required for his/her current development activity. This *context-dependent provisioning* requires identification of (a) the current development activity a user is occupied with, (b) the role the user is currently conducting, and (c) the information and knowledge objects required for this activity. The latter part becomes known as the result of the previously conducted analysis of the knowledge characteristics of the various development activities of MPD. At all times, a user will be able to indicate which development activity he/she is occupied with. Based on this indication, the design support system will provide the user with pre-built queries to determine the information and knowledge objects of interest.

Requirement 8.5: Furthermore, the networks of interlinked information can be queried to infer new insights from the wealth of established models in the context of the current product development activity. In cases where solutions of previous products or product iterations are of interest, the *queries* can be extended to past products.

Requirement 8.6: For each individual, user specific *knowledge profiles* can be established, which capture the depth and width of competences. When presenting information and knowledge to these actors, the knowledge profiles can be employed to control the width and depth of available information. For specialists in a certain domain, for example, the full range of available information can be presented, whereas users from other domains will be presented only a view of the information understandable to them.

The identified process elements for MPD described in appendix 1 give a good orientation on the use cases to be covered. The various actors involved in the Mechatronic product development process have been already compiled in appendix 2.

8.2 Semantic Technologies for the Semantic Web

The *Semantic Web* describes both the vision and the technological concept for the evolution of the current World Wide Web toward a web of linked data through the semantic enrichment of the current webpages (Berners-Lee, Hendler, et al. 2001). For this purpose, a set of semantic technologies provides a semantic layer as *"glue"* between the webpages and the potential consumers (Kiefer and Bernstein 2011). The semantic web stack describes this collection of technologies (and their layering), aiming at realizing the vision of the *Semantic Web* (cf. Figure 8.1).

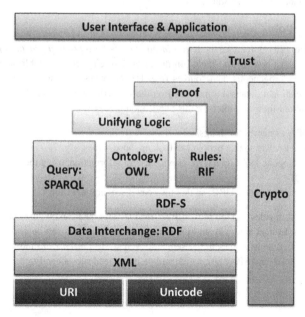

Figure 8.1: Semantic web stack, adapted from (Bratt 2007)

Unicode constitutes the preferred encoding schema for text data in the web context, which is currently in use for three quarters of all web sites. *XML* adopts *Unicode* as its prevailing character encoding schema and represents above all an option for the realization of the syntactical layer of the *Semantic Web*. Besides *XML*, alternative syntaxes for the *Semantic Web* exist, e.g. *Turtle* that groups three *URIs* into a very compact triple representation.

In very general terms, a uniform resource identifier (*URI*) can be perceived as a unique name to identify a resource. It can either take the form of a uniform resource locator (*URL*), of a uniform resource name (*URN*), or of both. In terms of the web, a *URL*[36] describes (a) the location of a webpage or other documents reachable from the web and (b) the protocol (e.g. HTTP, FTP, …) used in the connection. In contrast, a *URN* identifies a resource in a given namespace (e.g. ISBN) by a namespace specific string (e.g. the ISBN number). All of the three aforementioned concepts and technological standards (*Unicode*, *URI*, and *XML*) can be considered as foundations of the current web and represent therefore no distinctive features of the *Semantic Web*.

In the essence, the *Semantic Web* constitutes a concrete specialization of a semantic network[37] and an innovation driver for semantic technologies (Conrad 2010). The *Semantic Web*, however, employs a strongly decentralized and loosely coupled architecture that is clearly distinct from the centralized knowledge bases employed in traditional approaches to semantic networks. Overall, the *Semantic Web* approach does not intend to teach computers to infer the meaning of the published web pages, instead data and meaning will be expressed in a machine-understandable format (Bratt 2007).

8.2.1 Linked Data

Heath and Bizer (2011) describe the term *Linked Data* as *"a set of best practices for publishing and interlinking data on the Web."* In the essence, it enables the establishment of typed links between heterogeneous data sources (e.g. HTML pages, text documents, spreadsheet data, data in relational databases) published on the web (Bizer, Heath, et al. 2009). Berners-Lee (2006) laid out four principles for this publication process:

(1) Use URIs as names for things.
(2) Use HTTP URIs so that people can look up those names.
(3) When someone looks up a URI, provide useful information, using the standards (RDF, SPARQL).
(4) Include links to other URIs, so that they can discover more things.

Based on these four rules and the proven *Web Architecture*, data can be published on the web in a machine-understandable way and links to other data resources can be established. From the aforementioned technologies in the semantic web stack, *Linked Data* introduces the new technologies of RDF and SPARQL, which will be introduced subsequently.

---

[36] In many cases, the terms URI and URL are used in a synonymous way. The W3C provides some clarifications and recommendations under: http://www.w3.org/TR/uri-clarification/

[37] Semantic networks constitute a network of concepts linked through associations (Alwert and Hoffmann 2003).

8.2.1.1 RDF

The Resource Description Framework[38] (RDF) constitutes the language for representing *Linked Data* within the *Semantic Web*. RDF is typically applied to provide metadata about a resource on the web using a simple data model that comprises:

- The *subject* uses a URI to describe the resource that the metadata is provided for.
- The *predicate* describes the type of relationship (e.g. *"shall have as part a"*, *"has been replaced by"*) between the subject and object. The predicate refers to defined types of relationships in *vocabularies* (cf. section 8.2.2) by an URI.
- The *object* contains the described characteristic either as a literal value (e.g. string, number, date) or as an URI identifying the resource related to the subject through the predicate relationship.

Figure 8.2 illustrates the approach taken by RDF to provide additional information for a resource. In the chosen example, the RDF statement describes the fact that the electric motor shall have a hall sensor as component. In this case, each part of the RDF triple points to resources by means of an URI. The URIs of the subject and object identify the motor and the sensor, whereas the URI of the predicate refers to a specific relationship defined in a vocabulary.

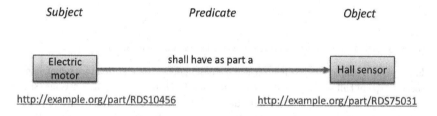

Figure 8.2: RDF describing the usage of a hall sensor as component of an electric motor

In this example, the URIs all point to the same website. Beyond that, RDF enables mixing of URIs from different websites in a heterogeneous manner. Several RDF triples can be merged into a labeled, directed graph of vertices (subjects and objects) and edges (predicates). Within the resulting single graph, predicates originating from different vocabularies can be mixed. Overall, such graph resulting from the merging of multiple RDF triples constitutes a semantic network.

Data models in RDF can be serialized in different syntactic formats. For two of these formats, RDF/XML and RDFa, dedicated W3C standards are available, whereas a candidate recommendation has been submitted for the popular Turtle syntax (Terse RDF Triple Language).

---

[38] The W3C provides access to the standards and drafts for RDF under: http://www.w3.org/standards/techs/rdf

In the context of semantic technologies, RDF became a de facto standard for providing metadata, although RDF is not an optimal modeling language for this purpose and is for instance impeded by limitations such as that the predicate can only refer to a single object (GEIST Research Group 2011).

8.2.1.2 SPARQL

SPARQL[39] constitutes a RDF query language, a protocol for communicating these queries and a result data format. It aims at querying RDF graphs and applies simple RDF graphs as queries for this purpose. Figure 8.3 depicts an example for the syntax of a typical SPARQL query: Within the PREFIX section, namespace prefixes (e.g. ex, rdf) can be declared that help to use a shortened notation when writing the queries.

```
PREFIX ex:  <http://example.org/schema#>
PREFIX rdf: <http://www.w3.org/1999/02/22-rdf-syntax-ns#>
SELECT ?partNumber ?description
WHERE
{
        ?electricMotor rdf:type ex:ElectricMotor .
        ?electricMotor ex:shallHaveAsPartA
            <http://example.org/part/RDS75031> .
        ?electricMotor ex:partNumber ?partNumber .
        ?electricMotor ex:description ?description .
}
```

Figure 8.3:   SPARQL query to find all electric motors containing a particular hall sensor

In the given example, the part number and the description of all electric motors containing a specific type of hall sensor should be returned as result. Therefore, the SELECT clause specifies two variables ?partNumber and ?description that describe the content of the results. The WHERE clause contains the triple patterns to match enclosed in curly braces. Within the given example, the variable identifier ?electricMotor stands for the electric motor with the two specified properties.

The SPARQL specification offers additional syntactic flavors for the WHERE clause:

- The OPTIONAL key-word allows specifying patterns not occurring in all results. If these patterns are matched, they will contribute additional data to the result.
- The UNION key-word permits specification of alternative patterns that will be joined together in the result.
- The FILTER key-word restricts the results to those passing the filter condition.
- In addition, SPARQL provides various operators e.g. for comparison, testing, and casting.

---

[39] The W3C provides access to the specifications for SPARQL 1.1 under: http://www.w3.org/TR/sparql11-overview/

So far, only the `SELECT` clause was applied for specifying a table representation of the results. Other key words exist to specify alternative result formats:

* The `CONSTRUCT` key word allows specifying a RDF graph as result format.
* The `ASK` key word specifies queries returning only "true" or "false" depending on whether the query was matched.
* The `DESCRIBE` form returns as result an RDF graph that comprises RDF data about the resource, where the description is determined by the query service.

As previously mentioned, the SPARQL specification defines also a protocol describing the transmission of a query to a SPARQL processing service and how the results are returned to the requester. Within the SPARQL 1.1 specification, the W3C added support for federated SPARQL queries encompassing multiple endpoints.

The essential features of SPARQL as query language for RDF graphs have been introduced. SPARQL does not provide, however, direct support for neither RDF Schema nor OWL (Hitzler, Krotzsch, et al. 2011).

8.2.2 Vocabularies

Within the vision for the *Semantic Web*, URIs, RDF, and SPARQL provide the foundational aspects of *Linked Data*, whereas vocabularies contribute the semantics to overall vision. In simple cases, the used vocabulary does not need to be declared explicitly – instead it will be coded into the particular application. For data integration purposes, lightweight vocabularies might be appropriate that allow only for limited reasoning scenarios. For other purposes, more elaborate reasoning capabilities might be required that presuppose complex ontologies. Overall, the term *"reasoning"* in the context of the Semantic Web refers to the discovery of new relationship from the used ontologies.

The *Semantic Web* initiative does not intend that all vocabularies will be consolidated in a single ontology. Instead, it favors a bottom-up movement where domain–specific communities agree on the vocabularies in their domain. Wherever possible, terms from existing vocabularies should be reused. This approach is well-suited to the decentralized architecture of the *Semantic Web*.

When introducing RDF and SPARQL in the previous sections, vocabularies[40] have been mentioned as the source of specific terms used in RDF triples. The traits and the purpose of such vocabularies will be introduced subsequently.

8.2.2.1 RDF Schema

RDF Schema[41] (RDF-S) allows specifying lightweight ontologies in RDF that typically consist of classes and properties as well as their hierarchies. Moreover, RDF-S supports the following conceptual features:

---

[40] In the following, the term *"vocabulary"* will be used as an umbrella term describing taxonomies, vocabularies, and ontologies all allowing for a definition of the concepts, their relationships, and constraints within a specific domain.

- Definition of classes and specification of resources as instances of such classes.
- Using the predicate rdfs:subClassOf one class can be declared a subclass of a base class. Consequently, class hierarchies can be designed.
- Using the predicate rdfs:subPropertyOf specializations of properties can be declared, which allows for the design of property hierarchies.
- The predicate rdfs:domain allows classifying the type of the subject for a property.
- The predicate rdfs:range permits to define the type of object for a property.

Consequently, the rdfs:domain predicate permits to declare a property as part of the class of its subject.

Due to its simplicity, RDF-S is considered as the most widely applied ontology language today (Stuckenschmidt 2009). It permits to describe rather uncomplex ontologies together with the associated data.

8.2.2.2 OWL

The *Semantic Web* positions the Web Ontology Language[42] (OWL) for the definition of more complex ontologies. OWL builds upon RDF and RDF-S and provides the vocabulary required for the following use cases:

- When defining synonyms or mappings between different vocabularies, terms can be defined as equivalent:
  - o Using the predicate owl:equivalentClass one class can be declared as equivalent to another one.
  - o Using the predicate owl:equivalentProperty one property can be declared as equivalent to another one.
- Likewise, terms can be defined as mutually exclusive, i.e. an instance can either be member of one term, but not of both:
  - o Using the predicate owl:disjointWith one class can be declared as disjoint to another one.
  - o Using the predicate owl:disjointPropertyWith one property can be declared as disjoint to another one.
- At the level of properties, the cardinality of the relationship can be defined. In addition, it is possible to define a range by using the pair of predicates owl:minCardinality and owl:maxCardinality.
- In addition, using the predicate owl:inverseOf a property can be defined as the opposite of another property.
- At the level of individuals, the predicate owl:sameAs describes the adherence of two resources to a common concept.

---

[41] The W3C provides access to the specification for RDF-S under: http://www.w3.org/TR/rdf-schema/

[42] The W3C provides access to the specifications for OWL under:
http://www.w3.org/standards/techs/owl#w3c_all

- Classes can be defined from enumerated values.

The aforementioned listing describes a set of commonly used features of OWL. However, OWL contains various other features, e.g. to further characterize properties as symmetric, transitive, reflexive, etc. If more of these advanced features of OWL are used, then the implementation of the inference engines coping with such features becomes rather complex. Therefore, OWL offers three official sublanguages (cf. Figure 8.4) with the following characteristics (Hitzler, Krotzsch, et al. 2011):

- *OWL Full* covers all OWL language features in conjunction with RDF-S. Due to the resulting complexity, there is currently no inference engine capable of supporting the complete semantics of OWL Full. Therefore, OWL Full is mainly used for conceptual modeling.
- *OWL DL* supports all OWL language features, but some of them with certain restrictions. These restrictions aim at allowing reasoning by inference engines.
- *OWL Lite* imposes further restrictions on top of the restrictions of OWL DL in order to become an easy to implement OWL sublanguage. The implementation of inference engines for OWL Lite, however, turned out to be almost as difficult as for OWL DL. Therefore, OWL Lite only plays a marginal role.

Overall, OWL DL is the most important of the presented sublanguages (Hitzler, Krotzsch, et al. 2011). It represents a description logic[43], hence the name DL. In addition, the OWL 2 standard offers three language profiles (OWL 2 EL, OWL 2 QL, OWL 2 RL) selected for their computational characteristics.

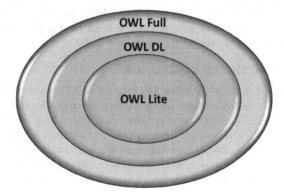

Figure 8.4: OWL sublanguages

Besides the already introduced semantic technologies for *Linked Data* and the definition of *vocabularies*, the semantic web stack (cf. Figure 8.1) contains other technologies (e.g. the Rule Interchange Format), which are not in the focus of the present thesis.

---

[43] The term *description logics* describe a family of knowledge representation formalisms that are typically fragments of first-order predicate logic (Hitzler, Krotzsch, et al. 2011).

8.3 Semantic Technologies for Design Support in MPD

As previously stated, access to the information stored in the various types of models presents essentially a *data integration* problem. Due to the various languages and metadata models typically used, the approach for the data integration of models in MPD will have to address a range of syntactic and semantic problems. Stuckenschmidt (2009) structures such issues for the integration of data from heterogeneous sources in three categories:

(a) *Syntactic conflicts* arise from differences in the used languages and data models. As analyzed in section 2.6.3, more than 30 types of models are used throughout the different design phases of MPD. Depending on the concrete toolset, a few of these models may share common metadata models (e.g. SysML, STEP) and may be formatted in the same language (e.g. XML). It is, however, unrealistic to assume that all of the employed models in MPD can adhere to the same metadata model and be saved in the same language. Therefore, the previously introduced *Linked Data* approach will be adopted for coping with syntactic conflicts: In addition to the native model formats, the data of the models will be *published* and *interlinked* based on *RDF* and *metadata* originating from *shared vocabularies*.

(b) *Structural conflicts* are induced by structures differing between data sources: Here, the different data sources may keep the same data using attributes with different names or data types. Likewise, the same data can be dispersed over differing number of objects.

(c) *Semantic conflicts* result from the usage of different terms for identical concepts or predicates.

For the latter two problems, ontologies constitute a well-established means of coping with data heterogeneity. They provide (a) a neutral data model for the heterogeneous data sources, (b) a complete specification of all involved concepts and implicit assumptions, and (c) inference methods to detect inconsistencies (Stuckenschmidt 2009).

8.3.1 Analysis and Selection of Vocabulary

The first requirement described in section 8.1 (Requirement 8.1) focuses on data integration for the native models through *metadata* originating from shared *vocabularies*. In this way, the *Linked Data* approach using RDF *metadata* originating from *shared vocabularies* will be adopted, as previously mentioned. This approach leaves, however, much freedom to choose the appropriate type of vocabulary according to the specific requirements of the scenarios to cover. In order to choose the most suitable type of vocabulary and implementations of vocabularies covering specific aspects, the following issues will have to be analyzed and discussed:

(1) What type of vocabulary will support the required data integration of models in MPD? Here, the following sub-issues will be considered:
   (a) What level of expressiveness is required?
   (b) Is support for inferencing required?

(2) Which subjects of conceptualization (i.e. application, domain, upper ontology, and knowledge-representation ontology) should be supported?

(3) What are the domains of MPD to be covered?

(4) Are there any implementations of vocabularies for the identified domains available?

Henrichs (2009) proposes a two-dimensional scheme for the classification of (a) the *expressiveness* of ontology and (b) the *subject* of the *conceptualization*. For the first dimension, he adopts the spectrum of ontologies described by (McGuinness 2005) that reaches from *controlled vocabulary* to *ontologies* with *inverse* and *disjointness* relationships[44]. The second dimension consists of four subject levels of conceptualization, describing the level of specialization of the ontology:

(a) The *knowledge representation ontology* (KR ontology) forms the most general level of conceptualization that typically provides generic modeling primitives (e.g. concept, relationship, attribute) only.

(b) An *upper* or *foundation ontology* defines concepts and relationships encountered in most domains.

(c) The *domain ontology* contains the domain-specific terms and relationships encountered in a domain as e.g. mechanics, electronics, and software. The domain ontology typically reuses terms defined in upper ontologies.

(d) The *application ontology* comprises the kinds of items and their relationships specific to a tool. It typically builds up on the concepts and predicates from the higher subject levels.

Usually, the lower subject levels depend on the concepts and relationships defined by higher levels.

The definition of MPD established in section 2.4.1 mentions four relevant engineering domains (mechanical engineering, electrical engineering, and computer science with control engineering as one of its sub-disciplines, and systems engineering), where systems engineering contributes to (a) the adopted methodologies and procedures, and defines (b) system-level terms adopted for the description of architectures, functions, etc.

Figure 8.5 depicts the resulting layering of ontologies from the different subject levels of conceptualization as conceived for the design support system of MPD.

Besides the KR ontology as meta-meta-language, the design support system will require the following subject levels of conceptualization (cf. Figure 8.5):

(a) The *foundation ontologies* comprise the concepts and their relationships at a general physical level (e.g. measures and units), and at the system-level (e.g. requirements, functions, system architecture). They use the modeling language defined by the selected KR ontology.

(b) The *domain ontologies* define the domain-specific terms and their relationships in e.g. mechanics, E/E, software, and for control systems. They reuse the terms defined by the foundation ontologies and apply the modeling language given by the KR ontology.

---

[44] Detailed descriptions of the various structures of conceptualizations can be found in (Henrichs 2009).

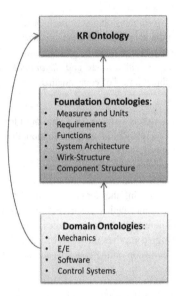

Figure 8.5: Hierarchy of ontologies for the design support system in MPD

The previously identified requirements do not indicate a need for the publication of models in dedicated application ontologies. If the need for such application-specific ontologies arises, however, they could be specified and integrated quickly based on the principles of *Linked Data*.

In the following, the focus of the discussion returns to the first dimension of the classification scheme, i.e. the required level of *expressiveness* to support the data integration of models in MPD. In order to decide on the type of ontology to be adopted, the first step will analyze the characteristics of a portfolio of modeling languages. In the second step, the required conceptual features (i.e. the structure of conceptualization) will be defined in order to select the most suitable language. In addition, the question of the support for inferencing has to be discussed.

Drawing from (Henrichs 2009), in addition to the previously presented modeling languages (XML, RDF, RDF-S, and the different sub-languages of OWL) the following two modeling languages will be included in the analysis:

(1) The *ISO standard 10303* (*STEP*) aims at enabling the exchange of product model data. It comprises several conceptual levels ranging from the modeling language for product data (EXPRESS), over shared foundation resources (ISO 10303-4x) to concrete application protocols for specific domains and design phases (ISO 10303-2xx). As the EXPRESS language originated in the 1980's, it is nowadays considered a legacy that only a few specialists and tools employ. Therefore, it needs to be replaced by a state-of-the-art means of information modeling which allows for the description of the STEP data models in ontology languages. For this purpose, the Object Management Group (OMG) has specified the mapping of the EXPRESS

metamodel to the UML 2 and OWL metamodels.[45] Consequently, the ISO standard models described in EXPRESS can be translated into UML models and OWL ontologies.

(2) van Renssen (2005) conceived *Gellish* (originally for "Generic Engineering Language") as a formal generic artificial language that can be employed for an unambiguous and machine-interpretable description of technical artifacts. It constitutes a subset of natural languages in the variants *Gellish English* and *Gellish Nederlands*, where, based on a dictionary expressions in one language variant can be translated into the other one. As indicated by Figure 8.6, *Gellish* covers the four subject levels of conceptualization. The language itself is based on a similar grammar to RDF: Facts are described as triples of the type *Object$_1$-Relation-Object$_2$*. The upper ontology (called the *TOPini* table) consists of more than 1500 concepts and 650 relationship types, while the dictionary/taxonomy[46] contains more than 40000 concepts. These concepts and relationships originate from the EPISTLE core model and several STEP application protocols, e.g. AP 221, AP 226, and AP 231. In addition to the description of facts, *Gellish* equally allows expressing queries and messages.

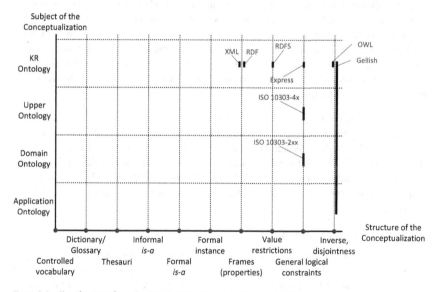

Figure 8.6: Classification of ontology modeling languages, adapted from (Henrichs 2009)

Figure 8.6 indicates that Gellish and OWL achieve the highest level of *expressiveness* whereas the EXPRESS language does not offer relationships to describe disjointness and inversion out of the box. Regarding the achieved subject levels of conceptualization, Gellish covers all

---

[45] The OMG provides access to the EXPRESS metamodel under: http://www.omg.org/spec/EXPRESS/1.0/

[46] The *Gellish* dictionary/taxonomy is available from http://sourceforge.net/projects/gellish/

four levels, whereas OWL covers as language only the KR level. Based on EXPRESS, the various levels of the ISO standard 10303 (STEP) cover the upper three subject levels of conceptualization.

For the intended design support system of MPD, several advanced conceptual features are required, e.g. cardinality, equivalence to tackle difference in naming between domains, and enumerations. The conceptual features placed at the very right of the classification of ontology modeling languages depicted in Figure 8.6, i.e. disjointness and inversion, are not required in the context of the design support system. In conclusion, all three languages achieve the required expressiveness.

Given the current state of mandatory requirements, inferencing is not required in the frame of the intended design support system. However, the need to apply inferencing may easily arise (a) for the detection of inconsistencies between ontologies (Stuckenschmidt 2009) and (b) in use cases where the mapping of measures and units is required (Rijgersberg, van Assem, et al. 2013). Neither EXPRESS nor Gellish provide support for inferencing, whereas inference engines for OWL-Lite and OWL-DL are available.

Although all three languages (EXPRESS, Gellish, OWL) provide the required expressiveness, and EXPRESS and Gellish support vocabularies on at least two additional subject levels, both languages are (a) difficult to integrate with the approaches of the Semantic Web and (b) not sufficiently scalable for technologies like inferencing that might be required at some point in time. Their advantage of providing established vocabularies, however, is relatively unimportant as these vocabularies can be "harvested" and translated into OWL due to the similar metamodels of the three languages.

For these reasons, OWL-DL will be selected as the ontology modeling language. In order to establish the required vocabularies indicated in Figure 8.5, the existing ISO standard models described in EXPRESS can be translated into OWL using the EXPRESS metamodel specified by the OMG. The STEP ISO standard offers support for measures and units, system architecture (AP 233), mechanical design (AP 214), and E/E design (AP 210, 212). For other areas such as Wirk-structure, software, and control systems, supplementary ontologies will have to be identified or developed.

8.3.2 Interlinking of Models

The principles of *Linked Data* described in section 8.2.1 will be adopted for the publishing of the resources contained in the models (cf. Requirement 8.2). Consequently, each resource will be described by an URI representing the resource as unique identifier. Based on this mechanism, the resource can be used as subjects and objects in RDF triples providing the metadata on the models. The same mechanism can be equally applied for the interlinking of resources inside of one model and between different models. Consequently, networks of interlinked data between the models will be established.

8.3.3 Semantically-Enabled Applications

The shared vocabularies indicated in Figure 8.5 can be adopted by applications to interpret the content of the RDF triples provided for the published models. Consequently, these applications are enabled to process the information and knowledge contained in the models based on the adopted vocabularies (cf. Requirement 8.3).

8.3.4 Queries

Using SPARQL, the networks of interlinked RDF triples can be queried to determine the information and knowledge objects of interest for a particular development activity (cf. Requirement 8.4), In addition, SPARQL can be applied to infer new insights from the published models (cf. Requirement 8.5). Firstly, models belonging to the current product development activity can be queried. Secondly, published models of previous product development activities can also be queried in order to gain information and knowledge on established solutions.

8.3.5 Knowledge-Profiles

All concepts and predicates published in RDF triples belong to the shared vocabularies depicted in Figure 8.5. For each piece of information described by a RDF triple, it can be therefore inferred whether it describes system-level information (e.g. by using the System Architecture ontology) or disciplinary information (e.g. by using domain ontologies). Based on this characterization according to the used ontology, the presented facts can be filtered to the *knowledge profile* of an actor (cf. Requirement 8.6). For this filtering process, the knowledge profile captures the depth and width of knowledge of an actor.

8.3.6 Conclusions

Altogether, the selected semantic technologies and the chosen vocabularies fulfill all of the requirements identified in section 8.1. Moreover, as a whole they constitute the answer to the fifth research question formulated in section 1.2.

9 Architecture for a Design Support System in Mechatronic Product Development

> *The world surrounding knowledge-rich systems changed drastically with the advent of the Web [...]. Knowledge-rich systems today are distributed, have many users with different degrees of expertise, and integrate many shared knowledge sources of varying quality.*

Gil, 2011

In the above quote, Gil (2011) emphasizes the fundamentally changed characteristics of knowledge-rich systems through the emergence of the Web and of *Semantic Web* technologies. In contrast to earlier knowledge-based systems that were depending on a central knowledge base, today's knowledge-based systems comprise a set of distributed knowledge bases usually with heterogeneous content. Likewise, the number of users providing content to the knowledge bases significantly increased from the very few users involved in earlier knowledge acquisition activities to the high number of users contributing their content in a Web-based setting. Altogether, the given description of today's knowledge-rich systems matches well the characteristics encountered in MPD, where many users with distinct competences contribute different pieces of information and knowledge in a distributed manner. Accordingly, the chosen set of semantic technologies for the design support system in the previous chapter matches well the distributed and heterogeneous style of activities encountered in MPD.

In the next step, an appropriate system architecture for the intended design support system will be developed that constitutes the answer to the sixth research question:

6. *What is an appropriate system architecture for a design support system based on semantic technologies aiming at improving the processing, storage, distribution, and context-sensitive provisioning of knowledge within MPD?*

The needed input for this task is provided by the identified high-level requirements of section 8.1, whereas the selected semantic technologies constitute a set of technical constraints for design and implementation. From the given requirements, the various functions to be realized by the design support system can be derived. In a subsequent step, a set of logical components supporting these functions will be identified. Interfaces between these components as well as communication standards will then have to be established. The technical constraints resulting from the chosen semantic technologies influence both the technical characteristics of the components and of their interfaces.

In the first step, section 9.1 applies use-case modeling to identify and describe the use cases targeted by the design support system, the involved types of users, and the boundaries between users and the systems or sub-systems. Next, section 9.2 identifies the necessary logical components and their interfaces for the realization of the functions of the design support system. Section 9.3 describes the conceived system architecture and provides the rationale that leaded to this architecture. Finally, section 9.4 evaluates the accomplished requirements coverage and summarizes the achievements of the system architecture for the design support system.

9.1 Use-Case Modeling

In the following, use-case modeling will be applied to identify and describe the main use cases targeted by the design support system. Each of the use cases consists of the involved actors (e.g. System Architect, Mechanical Design Engineer), the activities conducted by these actors, the system or sub-system used within these activities, and the relationships between individual activities and between activities and actors. The characteristics of each use case will be modeled in a use case diagram according to the Unified Modeling Language[47] (UML).

The process elements proposed in section 7.2.3 cover the inter-disciplinary and disciplinary development activities conducted in the frame of MPD. Within these development activities, a range of modeling and data management tools is already applied. The envisioned design support system, in contrast, intends to improve the creation, storage, distribution, and application of information and knowledge in this context. Clearly, it does not intend to replace the already utilized tools. Instead, it will provide merely integrative features not offered by the existing tool world, for which the use cases will be identified in the following. Use cases that focus on the authoring of the used vocabularies will be assumed as earlier development activities and are not included in the following use cases.

9.1.1 Use Case: Authenticate User

In order to load properly the associated knowledge profile, any user of the design support system needs to be authenticated. As part of the login process, the authentication of the user will be requested based on the user name and password. Once a user has been properly authenticated, the user's knowledge profile can be retrieved and used. This use case will be included in all of the following used cases.

9.1.2 Use Case: Filter the Retrieved RDF Triples by the Knowledge Profile of a User

The knowledge profile of a user captures the depth and width of knowledge of an actor. It will be applied for the automatic filtering of the retrieved RDF triples in several of the following use cases.

9.1.3 Use Case: Publication of URIs and RDF Triples for a Model

The various users authoring and releasing models (e.g. Mechanical design engineer, E/E design engineer, and Software developer) will be subsumed under the *Model Author* role that is illustrated by Figure 9.1.

---

[47] The OMG provides access to the UML specifications under: http://www.omg.org/spec/UML/

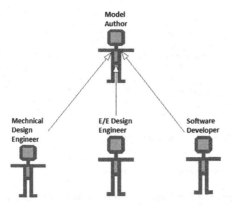

Figure 9.1: Users acting as Model Author

The authoring and releasing of a new version of a model by the *Model Author* triggers an automated process that generates and publishes the HTTP URIs and the RDF triples for the resources contained in the model (cf. Figure 9.2). Based on the vocabularies relevant for the particular model, the RDF triples will contain (i) relationships between the resources contained in the particular model and (ii) links to resources contained in other models.

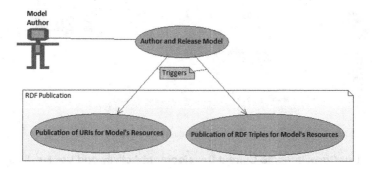

Figure 9.2: Use case - publication of URIs and RDF triples for the resources of a model

9.1.4 Use Case: Access the Resources and RDF Triples of a Model

For each released model, the information and knowledge contained in the model's resources should be directly accessible to the various actors involved in the Mechatronic product development process. The dedicated role of a *Model Description Reader* comprises all users involved in the development process, who need to consult the descriptions of the model's resources. As shown by Figure 9.3, the *Model Description Reader* points an RDF-enabled client application to the URI of the model's main resource or the URIs of other resources contained in the model. Then, the associated server application will send the RDF triples to the client that finally visualizes their content to the requesting user.

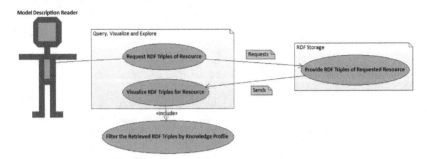

Figure 9.3:   Use case – access the resources and RDF triples of a model

9.1.5 Use Case: Explore the Structure of the RDF Triples of a Model

This use case focuses on the visualization of the structure formed by the links contained in the RDF triples. Initially, the *Model Description Reader* uses an RDF-enabled client application to retrieve the RDF triples associated with a model's resources, which are transformed subsequently into a structural representation. Then, the user can explore the visualized structure and navigate to resources linked to a model (cf. Figure 9.4). The characteristics of the visualized structure can be controlled by selecting the types of properties to be considered.

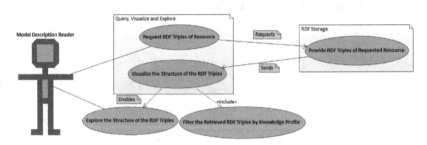

Figure 9.4:   Use case – explore the structure of the RDF triples of a model

9.1.6 Use Case: Indicate Current Development Activity and Use Pre-Built Queries

In the context of a SPARQL-enabled client application, the *Model Description Reader* can utilize a number of pre-built SPARQL queries to obtain the information and knowledge objects of interest. In the first step, the *Model Description Reader* has to indicate the development activity he/she is currently working on. Subsequently, the design support system will make the pre-built SPARQL queries available to the user, which can then be executed by the user.

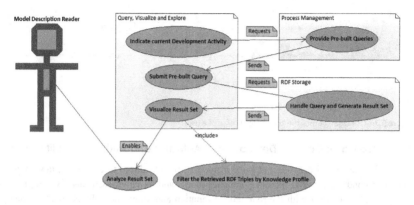

Figure 9.5: Use case – indicate the current development activity and use pre-built queries to determine the information and knowledge objects of interest

9.1.7 Use Case: Query the Metadata of Multiple Models and Discover New Information and Knowledge

Using a SPARQL-enabled client application, the *Model Description Reader* can author queries for the network of interlinked RDF triples formed by the metadata of published models. Subsequently, the client application visualizes the content of the obtained result set. Then, the user can explore and analyze the content of the result set and gain new insights from them (cf. Figure 9.6). In addition, the user can refine the used queries to adjust the obtained result set.

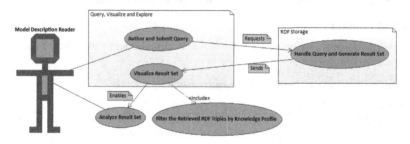

Figure 9.6: Use case – query the metadata of multiple models and discover new information and knowledge

9.1.8 Use Case: Author the Knowledge Profile of a User

The dedicated role of a *User Administrator* will be introduced in order to author and manage the knowledge profiles of the users (cf. Figure 9.7). For each user a competence profile can be defined, which refers to the shared vocabularies depicted in Figure 8.5. Either at the level of specific ontologies or specific properties it can be determined which types of facts are relevant for the user.

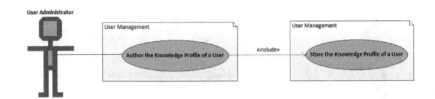

Figure 9.7: Use case - author the knowledge profile of a user

9.1.9 Use Case: Capture Development Activities and Author Pre-Built Queries

The dedicated role of a *Process Administrator* will be introduced to capture all development activities conducted within the development process and to author associated SPARQL queries providing the user with the required information and knowledge objects within a development activity (cf. Figure 9.8).

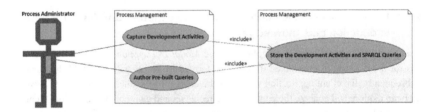

Figure 9.8: Use case – capture the development activities and author pre-built queries

9.1.10 Summary

Within the described use cases, three user roles were identified for which Table 9.1 shows the associated characteristics.

Table 9.1: Identified user roles for the design support system

Name	Description
Model Author	User authoring and releasing various types of models in the context of MPD. This user role subsumes the roles of e.g. the *Mechanical Design Engineer*, the *E/E Design Engineer*, and the *Software Developer*.
Model Description Reader	User consulting the descriptions of the model's resources, exploring the models' structure and authoring SPARQL queries on the base of existing metadata of models. This role potentially comprises all users involved in the development process.
User Administrator	User in charge of managing the roles and knowledge profiles of users.
Process Administrator	User in charge of managing the development activities and the associated SPARQL queries.

In the next step, the use cases were analyzed in order to identify the system-level and subsystem-level functionality used within these activities. At the system level, five constituents were found: The *Query, Visualize, and Explore* functionality comprises the functionalities for the authoring of RDF and SPARQL queries, the filtering, visualization and exploration of the obtained results. As indicated by Table 9.2, each of the modeling tools involved in the development process needs to support the *RDF Publication* for the native model data, based on the vocabularies relevant for the particular tool. Within the *RDF Storage* functionality of the design support system, the published RDF triples are stored, and requests for resources identified by HTTP URIs need to be handled. In addition, the *RDF Storage* implements a SPARQL endpoint for running SPARQL queries on specific parts of the published data. The *User Management* provides services for authentication, and the authoring and accessing of the knowledge profiles. Finally, the *Process Management* provides services for the capturing of the development activities, the authoring of pre-built SPARQL queries, and the accessing of pre-built SPARQL queries per development activity.

Table 9.2:   Identified functions for the design support system

System Functionality	Sub-system Functionality
Query, Visualize and Explore	Request and Visualize RDF Triples
	Query and Visualize Structure of RDF Triples
	Indicate current Development Activity and Submit Pre-built SPARQL Queries
	Author SPARQL Queries and Visualize Result Sets
	RDF Filtering
RDF Publication	*Remark: Required for each involved modeling tool*
RDF Storage	RDF Storage
	SPARQL Endpoint
User Management	User Authentication
	Knowledge-Profile Authoring
	Knowledge-Profile Storage
Process Management	Capture Development Activities
	Authoring of Pre-built SPARQL Queries
	Storage of Development Activities and SPARQL Queries

9.2 Architecture Development

In the next step, logical components for the realization of the functions specified in the previous section will be determined. This activity constitutes a functional partitioning of the identified functions into logical components. Subsequently, the architectural constraints guiding the design of the system architecture will be collected and further analyzed.

9.2.1 Functional Partitioning

By applying functional partitioning, the identified functions of the design support system and their associated data are mapped onto logical components. The resulting functional packages bundle several closely related functions and provide or receive data to or from other functional packages. Figure 9.9 depicts the functional system diagram that gives an overview on the identified logical components and their associated data. On the left side, it collects all functions relevant at *runtime* of the system, i.e. during the normal usage of the system by the *Model Author* publishing new versions of models or by the *Model Description Reader* querying and exploring the published metadata.

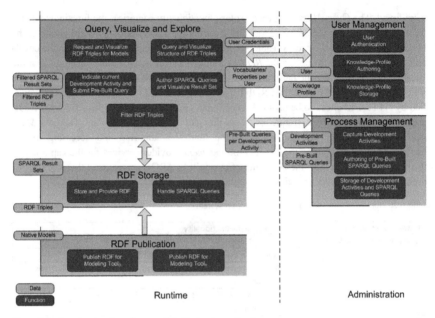

Figure 9.9:   Functional system diagram of the design support system

In contrast, the right side of the diagram depicts all administrative functions related to the management of users and processes. Arrows indicate the flows of data between the functional packages, e.g. the *RDF Publication* functionality submits RDF triples to the *RDF Storage*.

9.2.2 Architectural Constraints

For the subsequent development of the system architecture for the design support system, the following categories of architectural constraints will have to be taken into account:

- Interfacing of modeling tools for the publication of RDF
- Predetermined architectural principles and standards

9.2.2.1 Interfacing of Modeling Tools for the Publication of RDF

The majority of today's modeling tools is still implemented as traditional desktop applications (e.g. MATLAB/Simulink, most CAD tools), whereas the number of web-based modeling tools (e.g. CATIA V6) is only slowly increasing. Both types of modeling applications should be able to contribute the metadata of their models to the design support system, which requires an openness of the design support system for both architectures.

9.2.2.2 Predetermined Architectural Principles and Standards

The previously discussed adoption of the *Linked Data* principles (cf. section 8.2.1) implies the usage of certain standards for communication protocols and data persistency (see next section). Moreover, the design support system requires a distributed system architecture and the adoption of adequate technologies. Among the contemporary distributed system architectures, the following two architectures can be considered as the most influential ones:

(a) *SOAP*[48] (originally for Simple Object Access Protocol) specifies the protocols for a messaging framework that is widely used in the implementation of XML-based Web services. It provides the foundation of the *Service-oriented Architecture* (SOA), which describes an architectural style that is based on loosely coupled services, where each service contains the IT components implementing a business function.

(b) In contrast, Fielding (2000) proposes a more constrained architectural style called *Representational State Transfer* (REST) that likewise can be used for the implementation of Web services. It provides the basis for the *Resource-oriented Architecture* (ROA) and imposes the following set of additional restrictions compared to SOAP:

  i.   All resources are identified by *URIs*.

  ii. *HTTP methods* (e.g. GET, PUT, and DELETE) are applied for the manipulation of resources.

  iii. Within these actions, the *representation* of a resource carries the current or desired state of the resource between client and server.

  iv. The communication between client and server has to be *stateless*.

Obviously, several commonalities can be acknowledged between the *Linked Data* principles and the characteristics of *REST*, although both paradigms do not directly refer to each other. Overall, the *REST* architectural style supports all of the identified architectural constraints and will be adopted therefore for the design support system.

9.3 System Architecture of Design Support System

This section presents the system architecture as a whole and describes the rationale that led to this architecture. It lays out and justifies the decisions shaping the architecture concerning architectural constraints, selected technologies, and external components.

The design of the system architecture departs from the functional packages and the architectural constraints identified in the previous section. The architecture design process searches concrete software components for the realization of one or multiple functions contained in a functional package. Subsequently, the interfaces between these software components will have to be established based on communication standards compliant to the architectural constraints.

9.3.1 Differentiation from Standard Architectures and Standard Software

As specified in the previous section, the design support system requires a distributed system architecture based on the REST principles. Within this architecture, each functional package has to be associated either with the client side or with the server side. Obviously, certain functional packages (*RDF Storage, User Management, and Process Management*) have to be placed on the server side because these functions need to be accessible (a) from the many

---

[48] The W3C provides access to the SOAP specifications under: http://www.w3.org/TR/soap/

modeling applications publishing RDF triples for models and (b) from the user interface of the design support system.

Regarding topic (a), section 9.2.2.1 specified an identical architectural constraint stating that the interfacing of both desktop and web-based modeling applications is required. Due to the required support for desktop applications, modeling tools cannot generally appear as server components in the system architecture. This, however, is the overall assumption of the typical decentralized architecture applied for *Linked Data* and the *Semantic Web*. This decentralized architecture assumes that publishers of metadata act as server components, i.e. their resources can be reached by a URI and the publisher is able to handle requests for RDF resources on its own.

Because such decentralized architecture is not suitable here, an architecture centralized with respect to the *RDF Storage* and the handling of requests for RDF resources is required, i.e. both desktop and web-based modeling applications act as clients of the *RDF Storage* when publishing their RDF triples. The appropriate *RDF Publication* component for each modeling tool will have to be associated with the modeling application.

Recently, a draft for the upcoming W3C standard of the *Linked Data Platform*[49] was published. It provides a collection of best practices and an approach for a read-write *Linked Data* architecture. This to-be standard aims at achieving a deep integration between the models and tools by supporting a complete range of interoperability tasks like creation, updating, and deletion. This goes beyond the targeted *read and query scenarios* for the design support system. In addition, the *Linked Data Platform* solely supports the interoperability between web-based applications. Its adoption for the design support system would therefore require the wrapping of all modeling tools as desktop application by a web-based wrapper. Consequently, this architecture cannot be applied for the design support system.

Today, PDM systems manage a high portion of the documents and models used and generated during the product development process. However, the insight of the PDM system on the semantics embedded in the document is limited to the metadata extracted by a dedicated connector upon check-in. It typically remains at the level of administrative metadata for documents and at the level of the product structure information for CAx-models. Therefore, PDM systems do not support the consistent management for the whole of model entities contained in a model (Hahn, Gruening, et al. 2005). In order to enable a PDM system to comprehend the complete semantics of a model, both the data model of the PDM system and the tool-specific connector would have to be largely extended. Although this approach is technically feasible, it requires authoring access to the connectors of all involved tools. In addition, it will lead to a high number of tables in the underlying relational database of the PDM system, in order to capture the entirety of entities and their relationships in the various types of models. Both issues suggest that an extension of existing PDM systems in that direction is not advisable. Hahn, Gruening, et al. (2005) proposes therefore a new architecture of PDM systems based on semantic technologies. Once these novel PDM systems become commercially available, they

---

[49] The W3C provides access to the working draft for the Linked Data Platform under: http://www.w3.org/TR/ldp/

could provide the basis for the envisioned design support system. In the meantime, the design support system will however remain dissociated from today's PDM systems.

9.3.2 Separation of Client and Server Side Functionality

In the next step, the affiliation of a functional package to either the client or the server side will be specified. Table 9.3 summarizes the results of the partitioning along this dimension.

Table 9.3:　Partitioning of functional packages into client-side and server-side functionality

System Functionality	Client-side Functionality	Server-side Functionality
Query, Visualize and Explore	Request and Visualize RDF Triples	
	Query and Visualize Structure of RDF Triples	
	Indicate Current Development Activity and Submit Pre-Built SPARQL Queries	
	Author SPARQL Queries and Visualize Result Sets	
	RDF Filtering	
RDF Publication	*For desktop-based modeling tools*	*For web-based modeling tools*
RDF Storage		RDF Storage
		SPARQL Endpoint
User Management		User Authentication
	Knowledge-Profile Authoring	
		Knowledge-Profile Storage
Process Management	Capture Development Activities	
	Authoring of Pre-Built SPARQL Queries	
		Storage of Development Activities and SPARQL Queries

Obviously, the required user interface elements providing access to the *Query, Visualize and Explore* functionality will have to be placed at the client side. For desktop-based modeling tools, the *RDF Publication* functionality needs to be closely associated with the authoring tool and therefore will be placed at the client side. For web-based modeling tools, the same reasoning leads to the conclusion that the *RDF Publication* functionality has to be placed at the server side. As previously discussed, the functional packages of *RDF Storage, User Management,* and *Process Management* have to be placed generally on the server side with the exception of the user interfaces for the authoring of the knowledge profile, the capturing of the development activities and the authoring of pre-built SPARQL queries.

9.3.3 Logical and Physical System Architecture

Regrettably, there is no general agreement on the structuring elements used to describe a system architecture. Therefore, the well-established modeling of architectural elements as proposed by UML will be adopted to describe the system architecture by means of two types of UML diagrams:

(a) The *component diagram* describes the logical architecture. This visualizes a system's components, their interfaces, and the established connections between components.

(b) The *deployment diagram* captures the physical architecture. This describes the allocation of components to hardware nodes, the characteristics of the hardware and the standards used for the communication between the hardware nodes.

In the following, the identified functional packages will serve as the basis to introduce software components for their realization. In the first step, the previously conducted separation of client from server functionality permits the introduction of client and server applications for the identified client and server functionalities. This leads to the four top-level software components described in the first column of Table 9.4. Each top-level component comprises a range of associated sub-components captured by the second column of this table. The *Design Support Client Application* realizes the complete *Query, Visualize and Explore* functionality and the user interface for the *Knowledge-profile Authoring*. At the server side, the *Design Support Server*, the *User Management Server* and the *Process Management Server Application* will be considered as three distinct applications.

Table 9.4:   Identified top-level and sub-components of the design support system

Top-Level Component	Sub-Component
Design Support Client Application	Metadata Visualization User Interface
	Structure Visualization User Interface
	Pre-Built SPARQL Queries Submit User Interface
	SPARQL Query Authoring User Interface
	SPARQL Result Set Visualization User Interface
	RDF Filtering Service
	Knowledge-Profile Authoring User Interface
	Development Activity Capturing User Interface
	Pre-Built Authoring SPARQL Queries User Interface
RDF Publication Provider	*Remark: Required for each involved modeling tool*
Design Support Server Application	RDF Storage Service
	SPARQL Endpoint Service
User Management Server Application	Authentication Service
	Knowledge-Profile Service
Process Management Server Application	Development Activity Service
	Pre-Built SPARQL Management Service

Figure 9.10 depicts the logical system architecture of the design support system at the top level by means of the component diagram. A compass icon on this diagram indicates the north direction supporting the idea that data on this diagram typically flows from north to south. Following this idea, the components collected in the *Modeling Tools* layer provide the RDF triples serving as input for the *Design Support Server Application* belonging to the middle layer of the logical architecture. On its part, this server application serves as SPARQL endpoint and allows to access published resources through REST-style interfaces indicated by a 'T' in the ball of the interface descriptor. In addition, the middle layer contains the *Process Management Server Application* that publishes two interfaces for the access to development activities and pre-built SPARQL queries per development activity. Furthermore, it comprises the *User Management Server Application* that provides two interfaces for authentication and

the handling of knowledge profiles. The *Presentation* layer contains the *Design Support Client Application* utilizing six of the aforementioned interfaces.

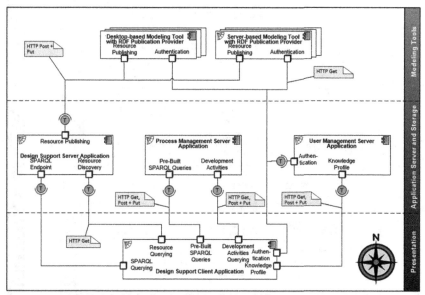

Figure 9.10: Component diagram describing the logical system architecture of the design support system

Appendix 5 compiles component diagrams for the logical system architecture at the detailed level of each component depicted in Figure 9.10.

In the last step, the physical architecture of the design support system is captured by means of the deployment diagram. The physical architecture describes the allocation of software components to hardware nodes as well as the characteristics of the hardware nodes. Figure 9.11 gives an overview on a possible physical architecture without giving detailed descriptions on (a) the used hardware, (b) used middleware and software frameworks and (c) precautions taken to assure scalability and fail-safety. It should merely be seen as a template for the physical architecture that can be further detailed once concrete requirements for the number of users, the targeted performance, policies on middleware to be used etc. become available.

In any case, the chosen physical architecture assumes that no access of clients from outside of the corporate network is required. Consequently, all client nodes are located in the *intranet* zone. In contrast, all of the server nodes are placed in the *core* zone.

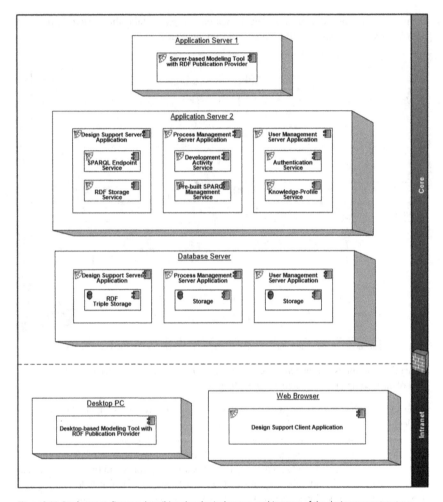

Figure 9.11: Deployment diagram describing the physical system architecture of the design support system

9.4 Assessment of Requirements Coverage and Conclusions

The described system architecture at the logical and physical level allows federation of the information and knowledge resources contained in the models published in the course of the development activities in MPD. Consequently, the design support system forms the single point of access to the resources of the models represented by the RDF triples. The developed system architecture conforms to the *REST* architectural style and respects all of the identified architectural constraints. As a whole, the proposed system architecture provides the contribution called for by the sixth research question expressed in section 1.2.

The system architecture allows realization of all of the use cases identified in section 9.1 as well as the functions captured by the use cases. Finally, it will be assessed to which degree the developed system architecture covers the high-level requirements established in section 8.1. For the assessment of the requirements coverage, the following symbols will be used:

O   =   no coverage

◑   =   partial coverage

●   =   full coverage

Table 9.5:   Assessment of the achieved requirements coverage by the proposed system architecture

ID	Requirements	Achieved Coverage
8.1	*Attribution of the native model data* by metadata originating from *discipline-specific vocabularies*	●
8.2	*Linking of the internal resources to external resources* based on common vocabularies	●
8.3	*Common vocabularies provide access to information and knowledge* embedded in models to all applications sharing the vocabularies.	●
8.4	*Context-sensitive provisioning* of the information and knowledge objects required for the current development activity	●
8.5	The *networks of interlinked information can be queried* to infer new insights from the wealth of established models in the context of the current product development activity	●
8.6	*Knowledge profiles* capture the depth and width of knowledge of each individual user.	●

10 Summary and Outlook

One of the major obstacles for the preparation of the present thesis arose from the "Babyloni-an confusion" on knowledge related terms (e.g. *tacit, implicit, explicit, declarative, procedural* knowledge). Often, these terms coexist in different areas of science with slightly different or even largely diverging meanings. As shown for the case of *implicit* knowledge in the context of information science, there is no consistent perception of this term in just one of the many scientific fields. Foremost, the present thesis had therefore to establish a consistent and coherent terminology for the concepts of data, information, knowledge encountered in humans and information systems that covers the scientific fields of cognitive psychology, information science, theories on organizational knowledge creation, design cognition, and interdisciplinary product development. The adoption of the model of organizational knowledge creation (Nonaka and Takeuchi 1995) implied a focusing on *tacit* and *explicit* knowledge, where *explicit* knowledge is considered synonymous to *declarative* knowledge as well as *tacit* knowledge synonymous to *procedural* knowledge. Moreover, an understanding of *explicit* knowledge was established that perceives *explicit* knowledge as existing in different representations in the *subjective* domain of human agents (mental representations) and the *objective* dimension of information systems (physical representations by means of language or writing). Consequently, *explicit* knowledge can be considered as interchangeable between individuals given that the respective individual possesses the knowledge resource required to interpret the conveyed knowledge. Likewise, an individual can articulate *explicit* knowledge and hereby transform it into its physical form represented by language or writing. For the context of *tacit* knowledge, the model of organizational knowledge creation (Nonaka and Takeuchi 1995) contributes the understanding of the conversions enabling the transfer of *tacit* knowledge.

Finally, Stanovich's tripartite model of mind originating from cognitive psychology, allows to associate *tacit* and *explicit* knowledge with specific cognitive systems: The *autonomous* mind conducts preattentive, cognitive activities of intuitive thinking and depends on tacit knowledge, whereas the *reflective* and *algorithmic* minds rely on *explicit* knowledge.

10.1 Summary

Based on the achieved understanding of MPD as an interdisciplinary activity, the question was investigated as to how the individual and organizational activities of knowledge creation performed in this process can be modeled and analyzed. For this purpose, the integrated descriptive model of knowledge creation was introduced as theoretical basis that allows the description of (a) the interplay of individual and collective processes in product development and (b) the sources of knowledge applied within these activities. It employs the framework of *cognitive* activities in design proposed by (Visser 2006a) for explaining the patterns of knowledge application and creation within *individuals*, whereas it adopts the model of organizational knowledge creation to describe collective processes from a *socio-cultural* perspective. For the integration of these approaches, the integrated descriptive model of knowledge creation in interdisciplinary product development follows an approach that merges the two models based of their common conceptual elements.

As a second part of the research framework aiming at the analysis of knowledge characteristics in product development, an analysis and modeling method was proposed that is able to capture the various knowledge conversion activities described by the integrated descriptive model of knowledge creation. As the result of a selection process between six analysis and modeling approaches, the KMDL method was identified as the most suitable approach described in the research literature. In order to fulfill the complete range of identified requirements, however, the KMDL method had to be extended by the following set of features:

(a) Means for modeling of knowledge resources moderating the knowledge conversion activity

(b) A modified modeling pattern employing *knowledge objects* to represent tacit and explicit knowledge in its mental representation, whereas *information objects* are used for describing articulated, explicit knowledge

Subsequently, the research framework was applied for the analysis of knowledge characteristics of MPD, whereby the development process was represented by common process elements compiled from various procedure models of MPD. As first steps of this analysis process, the KMDL process and activity views were established for a set of representative process elements. Based on these views, the knowledge characteristics of the process elements were extracted and documented. The proposed research framework for the analysis of knowledge characteristics in MPD proved capable of modeling and analyzing all knowledge conversion activities for the considered process elements.

The results obtained from the previous research steps were used within a design support system that aims at improving the creation, storage, distribution, and context-sensitive provisioning of information and knowledge throughout MPD. In particular, the previously determined characteristics of common process elements for MPD were included within the design support system as a ready-to-use library of common process elements that provides the basis for *context-dependent provisioning* of information and knowledge objects. Based on an indication of a user's current development activity and his/her role, the design support system provides the user with pre-built queries to determine the information and knowledge objects of interest. By means of common vocabularies (i.e. metadata models or shared conceptualizations), the information and knowledge embedded in models' data can be made accessible to applications sharing the same vocabularies.

In the last step, a system architecture for the design support system was developed that conforms to the *REST* architectural style. It allows the federation of the information and knowledge resources contained in the models published in the course of the development activities in MPD. Altogether, the semantic technologies selected for the design support system match well the distributed and heterogeneous style of activities in MPD, where many users with distinct competences contribute different pieces of information and knowledge. Furthermore, the vocabularies can act as true mechatronic engineering language beyond the tool-specific and discipline-specific languages. They facilitate the cross-disciplinary coordination and collaboration hindered by the differences in terminology between the involved disciplines and tools.

In the course of the described research, appropriate answers to all of the six research questions formulated in section 1.2 were supplied.

10.2 Potentials for Future Research

While introducing Stanovich's tripartite model of mind, the potential of this model to design research has already been mentioned. In particular, it could contribute to a better understanding of the interwoven application of rational thinking and expert intuition in cognitive activities of engineering design. Moreover, the model could provide the theoretical basis for gaining a deeper understanding of the importance of hypothetical reasoning and cognitive simulations for creativity in design.

In the course of the research in the present thesis, frequent decisions were taken to steer the research activities in one direction. Consequently, alternative directions were touched only briefly. Some of these alternative research directions could be worth a second look. Currently, the descriptive model of knowledge creation in interdisciplinary product development assumes that only individuals can be considered as cognitive entities. As explained in section 6.1.1, the concept of *"socially distributed cognition"* appears as an interesting alternative approach to explain social aspects of knowledge creation in product development. It presupposes, however, acknowledgement that social groups possess the characteristics of cognitive entities. Although Cook and Yanow (1993) identified three substantial problems for awarding social groups the status of cognitive entities, further research on alternative perceptions of *"socially distributed cognition"* might yield interesting insights.

Third, research on a new architecture of PDM systems based on semantic technologies seems to be a promising direction that could provide an alternative technical basis for the described design support system.

References

Aamodt, A.; Nygård, M.: Different roles and mutual dependencies of data, information, and knowledge - An AI perspective on their integration. In: *Data & Knowledge Engineering* 16 (1995), pp. 191-222.

Ahmed, S.; Hacker, P.; Wallace, K.: *The role of knowledge and experience in engineering design*. In Proceedings of the International Conference on Engineering Design - ICED 05, Melbourne, 2005.

Alavi, M.; Leidner, D.E.: Review: Knowledge management and knowledge management systems: Conceptual foundations and research issues. In: *MIS quarterly* 25 (2001), pp. 107-136.

Alciatore, D.: *Definitions of "Mechatronics"*. Retrieved 26.01.2011, from http://www. mechatronics.colostate.edu/definitions.html.

Alciatore, D.G.; Histand, M.B.: *Introduction to Mechatronics and Measurement Systems*. McGraw-Hill Companies,Inc., 2011.

Alwert, K.: *Die integrierte Wissensbewertung — ein prozessorientierter Ansatz*. In: Mertins, K.; Alwert, K.; Heisig, P. (Eds.), Wissensbilanzen, Springer Berlin Heidelberg, 2005, pp. 253-277.

Alwert, K.; Hoffmann, I.: *Knowledge Management Tools*. In: Mertins, K.; Heisig, P.; Vorbeck, J. (Eds.), Knowledge Management: concepts and best practices, Springer, Berlin Heidelberg New York, 2003.

Ameri, F.; Dutta, D.: Product lifecycle management: Closing the knowledge loops. In: *Computer-Aided Design and Applications* 2 (2005), pp. 577-590.

Andersson, M.: *Object-oriented modeling and simulation of hybrid systems*. Lund Institute of Technology, Department of Automatic Control, PhD Thesis. 1994

Andreasen, M.M.: Modelling—The Language of the Designer. In: *Journal of Engineering Design* 5 (1994), pp. 103-115.

Andreasen, M.M.; Hein, L.: *Integrated Product Development*. IFS (Publications): Bedford, 1987.

Angerbauer, R.; Buck, R.; Doll, U.; Hackel, M.: *aquimo: Ein Leitfaden für Maschinen- und Anlagenbauer - Adaptierbares Modellierungswerkzeug und Qualifizierungsprogramm für den*

Aufbau firmenspezifischer mechatronischer Engineeringprozesse. VDMA Verlag: Frankfurt/M., 2010.

Avgoustinov, N.: *Modelling in mechanical engineering and mechatronics: towards autonomous intelligent software models.* Springer: London, 2007.

Bebensee, T.; Helms, R.; Spruit, M.: *Exploring the Impact of Web 2.0 on Knowledge Management.* In: Boughzala, I.; Dudezert, A. (Eds.), Knowledge Management 2.0: Organizational Models and Enterprise Strategies, IGI Global, Hershey, PA, 2011, pp. 17-43.

Bellalouna, F.: *Integrationsplattform für eine interdisziplinäre Entwicklung mechatronischer Produkte.* Ruhr-Universität Bochum, Fakultät für Maschinenbau, PhD Thesis. 2009

Bennis, W.G.: *Why Leaders Can't Lead: The Unconscious Conspiracy Continues.* Jossey-Bass: San Francisco, 1997.

Berners-Lee, T. Linked Data - Design Issues. 2006. Available from: http://www.w3.org/DesignIssues/LinkedData.html.

Berners-Lee, T.; Hendler, J.; Lassila, O.: The semantic web. In: *Scientific american* 284 (2001), pp. 28-37.

Beynon-Davies, P.: Formated technology and informed action: The nature of information technology. In: *International Journal of Information Management* 29 (2009a), pp. 272-282.

Beynon-Davies, P.: Neolithic informatics: The nature of information. In: *International Journal of Information Management* 29 (2009b), pp. 3-14.

Beynon-Davies, P.: Significant threads: The nature of data. In: *International Journal of Information Management* 29 (2009c), pp. 170-188.

Bishop, R.H.: *Mechatronics : An introduction.* Taylor & Francis: Boca Raton, 2006.

Bizer, C.; Heath, T.; Berners-Lee, T.: Linked Data - The Story So Far. In: *International Journal on Semantic Web and Information Systems (IJSWIS)* 5 (2009), pp. 1-22.

Blessing, L.T.M.; Chakrabarti, A.: *DRM, a Design Research Methodology.* Springer: London, 2009.

Bludau, C.H.; Welp, E.G.: *Wissensbasierte Entwicklung mechatronischer Produkte.* In *Proceedings of the 12th Symposium "Design for X",* Neukirchen, 2001, Merkamm, H. ed., pp. 117-126.

Bodendorf, F.: *Daten- und Wissensmanagement.* Springer: Berlin Heidelberg, 2006.

Borutzky, W.: *Bond Graph Modelling of Engineering Systems: Theory, Applications and Software Support.* Springer: New York, 2011.

Boucher, M.; Houlihan, D., System Design: New Product Development for Mechatronics, Aberdeen Group, Boston, MA, 2008.

Bradley, D.: Mechatronics - More questions than answers. In: *Mechatronics* 20 (2010), pp. 827-841.

Brand, F.: *Transdisziplinarität - Voraussetzung für naturwissenschaftlichen und mathematischen Erkenntnisgewinn.* In: Brand, F.; Schaller, F.; Völker, H. (Eds.), Transdisziplinarität: Bestandsaufnahme und Perspektiven : Beiträge zur THESIS-Arbeitstagung im Oktober 2003 in Göttingen, Universitätsverlag, Göttingen, 2004.

Brand, F.; Schaller, F.; Völker, H.: *Transdisziplinarität: Bestandsaufnahme und Perspektiven : Beiträge zur THESIS-Arbeitstagung im Oktober 2003 in Göttingen.* Universitätsverlag: Göttingen, 2004.

Bratt, S., Semantic Web, and Other W3C Technologies to Watch, W3C, 2007.

Broenink, J.F.: Introduction to physical systems modelling with bond graphs. In: *SiE Whitebook on Simulation Methodologies* (1999).

Brose Fahrzeugteile, Mechatronische Systeme für Fahrzeugtüren und -sitze, Coburg, 2006.

Buur, J.: *A theoretical approach to mechatronics design.* Technical University of Denmark, Institute for Engineering Design, PhD Thesis. 1990

Buur, J.; Andreasen, M.M.: Design models in mechatronic product development. In: *Design Studies* 10 (1989), pp. 155-162.

Casella, F.: *Proceedings of the 7th International Modelica Conference.* Como, 2009.

Chami, M.; Seemüller, H.; Voos, H.: *A SysML-based integration framework for the engineering of mechatronic systems.* In *Proceedings of the 2010 IEEE/ASME International Conference on Mechatronics and Embedded Systems and Applications (MESA)*, Qingdao, ShanDong 2010 IEEE, pp. 245-250.

Chater, N.; Oaksford, M.: *Normative Systems: Logic, Probability, and Rational Choice.* In: Holyoak, K.J.; Morrison, R.G. (Eds.), The Oxford handbook of thinking and reasoning, Oxford University Press, New York, 2012, pp. 11-21.

Choi, B.; Lee, H.: Knowledge management strategy and its link to knowledge creation process. In: *Expert Systems with Applications* 23 (2002), pp. 173-187.

Choy, S.Y.; Lee, W.B.; Cheung, C.F.: *A systematic approach for knowledge audit analysis: Integration of knowledge inventory, mapping and knowledge flow analysis.* In *Proceedings of the I-KNOW '04*, Graz, Austria, 2004, pp. 69-77.

Comerford, R.: Mecha… what? In: *IEEE Spectrum* 31 (1994), pp. 46-49.

Conrad, J.: *Semantische Netze zur Erfassung und Verarbeitung von Informationen und Wissen in der Produktentwicklung.* Universität des Saarlandes, Lehrstuhl für Fertigungstechnik/ CAM, PhD Thesis. 2010

Cook, S.D.N.; Yanow, D.: Culture and organizational learning. In: *Journal of management inquiry* 2 (1993), pp. 373-390.

Daenzer, W.F.; Huber, F.: *Systems Engineering – Methodik und Praxis.* Verlag Industrielle Organisation: Zürich, 1994.

Dais, S.: Herausforderungen eines Automobilzulieferers. In: *Automobiltechnische Zeitschrift – ATZ* 106 (2004), pp. 18-20.

Dassault Systèmes: *Press Release: Dassault Systemes Launches V6R2010.* Retrieved 22.03.2011, from http://www.3ds.com/company/news-media/press-releases-detail/release/ dassault-systemes-launches-v6r2010/single/2131/?cHash=4eab9a5e29a3e603b56122a301e6f 07e.

Dassault Systèmes: *CATIA V6 Systems.* Retrieved 22.03.2011, from http://www.3ds.com/de/products/catia/portfolio/catia-v6/v6-portfolio/d/virtual-design/s/ systems-1/?cHash=c58cc85dabc7c8feb282a232f3f93a1a.

Davenport, T.H.; Prusak, L.: *Working knowledge: How organizations manage what they know.* Harvard Business Press: Cambridge, MA, 2000.

de Silva, C.W.: *Mechatronics: An integrated approach.* CRC Press: Boca Raton, 2005.

Deng, Q.-W.: *A Contribution to the Integration of Knowledge Management into Product Development.* Otto-von-Guericke-University, Lehrstuhl für Maschinenbauinformatik, PhD Thesis. 2007

Diehl, H.: *Systemorientierte Visualisierung disziplinübergreifender Entwicklungsabhängigkeiten mechatronischer Automobilsysteme.* Technische Universität München, Lehrstuhl für Produktentwicklung, PhD Thesis. 2009

Dienes, Z.; Perner, J.: A theory of implicit and explicit knowledge. In: *Behavioral and brain sciences* 22 (1999), pp. 735-808.

Eder, W.E.; Hosnedl, S.: *Design engineering: A manual for enhanced creativity.* CRC Press: Boca Raton, 2008.

Ehrlenspiel, K.: *Integrierte Produktentwicklung: Methoden für Prozessorganisation, Produkterstellung und Konstruktion.* Carl Hanser, München; Wien, 1995.

El-khoury, J.: *A Model Management and Integration Platform for Mechatronics Product Development.* Royal Institute of Technology Stockholm, School of Industrial Engineering and Management - Department of Machine Design, PhD Thesis. 2006

Eppinger, S.D.: Patterns of product development interactions. In: *Innovation in Manufacturing Systems and Technology (IMST);* (2002).

Eppler, M.J.: *Making knowledge visible through knowledge maps: concepts, elements, cases.* In: Holsapple, C.W. (Ed.), Handbook on knowledge management: Knowledge matters, Springer, Berlin Heidelberg, 2003, pp. 189-205.

Evans, J.S.B.: Dual-processing accounts of reasoning, judgment, and social cognition. In: *Annu. Rev. Psychol.* 59 (2008), pp. 255-278.

Evans, J.S.B.; Stanovich, K.E.: Dual-process theories of higher cognition advancing the debate. In: *Perspectives on Psychological Science* 8 (2013), pp. 223-241.

Eversheim, W.; Niemeyer, R.; Schernikau, J.; Zohm, F.: *Unternehmerische Chancen und Herausforderungen durch die Mechatronik in der Automobilzulieferindustrie, Materialien zur Automobilindustrie, Band 23.* VDA-Verlag: Frankfurt/M., 2000.

Felgen, L.: *Systemorientierte Qualitätssicherung für mechatronische Produkte.* Technische Universität München, Lehrstuhl für Produktentwicklung, PhD Thesis. 2007

Fielding, R.T.: *Architectural styles and the design of network-based software architectures.* University of California, PhD Thesis. 2000

Finger, S.; Dixon, J.R.: A review of research in mechanical engineering design. Part I: Descriptive, prescriptive, and computer-based models of design processes. In: *Research in Engineering Design* 1 (1989a), pp. 51-67.

Finger, S.; Dixon, J.R.: A review of research in mechanical engineering design. Part II: Representations, analysis, and design for the life cycle. In: *Research in Engineering Design* 1 (1989b), pp. 121-137.

Follmer, M.; Hehenberger, P.; Punz, S.; Rosen, R.; Zeman, K.: *Approach for the creation of mechatronic system models*. In *Proceedings of the International Conference on Engineering Design - ICED 11*, Copenhagen, 2011, pp. 258-267.

Follmer, M.; Hehenberger, P.; Punz, S.; Zeman, K.: *Using SysML in the Product Development Process of Mechatronic Systems*. In *Proceedings of the 11th International Design Conference DESIGN 2010*, Dubrovnik - Croatia, 2010, pp. 1513-1522.

Frank, U.: *Spezifikationstechnik zur Beschreibung der Prinziplösung selbstoptimierender Systeme*. Universität Paderborn, Heinz-Nixdorf-Institut, PhD Thesis. 2006

Freisleben, D.: *Gestaltung und Optimierung von Produktentwicklungsprozessen mit einem wissensbasierten Vorgehensmodell*. Otto-von-Guericke-University, Lehrstuhl für Maschinenbauinformatik, PhD Thesis. 2001

Frodeman, R.; Klein, J.T.; Mitcham, C.: *The Oxford Handbook of Interdisciplinarity*. Oxford University Press: Oxford, 2010.

Gausemeier, J.; Ebbesmeyer, P.; Kallmeyer, F.: *Produktinnovation - Strategische Planung und Entwicklung der Produkte von Morgen*. Hanser Verlag: München Wien, 2001.

Gausemeier, J.; Feldmann, K.: *Integrative Entwicklung räumlicher elektronischer Baugruppen*. Hanser Verlag: München Wien, 2006.

Gausemeier, J.; Kahl, S., Architecture and Design Methodology of Self-Optimizing Mechatronic Systems, in: Milella, A.; Di Paola, D.; Cicirelli, G. (Eds.), Mechatronic Systems, Simulation, Modeling and Control, InTech Open Access Publisher, Vucovar, 2010, pp. 255-282.

Gausemeier, J.; Möhringer, S.: *New VDI Guideline 2206 - A Flexible Procedure Model for the Design of Mechatronic Systems*. In *Proceedings of the International Conference on Engineering Design - ICED 03*, Stockholm, 2003.

GEIST Research Group. Semantic Web 3a - RDF. 2011. Available from: http://ai.ia.agh.edu.pl/wiki/_media/pl:dydaktyka:semantic_web:geist-semweb-rdf-annotated.pdf.

Gil, Y.: Interactive knowledge capture in the new millennium: how the Semantic Web changed everything. In: *The Knowledge Engineering Review* 26 (2011), pp. 45-51.

Goel, V.: *Sketches of Thought*. MIT Press: Cambridge, MA, 1995.

Goel, V.: Cognitive Neuroscience of Thinking. In: *Handbook of Neuroscience for the Behavioral Sciences* (2007).

Gotel, O.C.Z.; Finkelstein, A.C.W.: *An analysis of the requirements traceability problem*. In *Proceedings of the First International Conference on Requirements Engineering*, Colorado Springs, CO, 1994 IEEE, pp. 94-101.

Gourlay, S.: Conceptualizing Knowledge Creation: A Critique of Nonaka's Theory. In: *Journal of Management Studies* 43 (2006), pp. 1415-1436.

Gourova, E.; Antonova, A.; Todorova, Y.: Knowledge audit concepts, processes and practice. In: *WSEAS Transactions on Business and Economics* 6 (2009), pp. 605-619.

Grimheden, M.: *Mechatronics engineering education*. Royal Institute of Technology Stockholm, School of Industrial Engineering and Management, PhD Thesis. 2005

Grimheden, M.; Hanson, M.: *What is mechatronics? Proposing a didactical approach to Mechatronics*. In *Proceedings of the 1st Baltic Sea Workshop on Education in Mechatronics*, Kiel, 2001.

Grimheden, M.; Hanson, M.: Mechatronics - the evolution of an academic discipline in engineering education. In: *Mechatronics* 15 (2005), pp. 179-192.

Gronau, N.: *Modeling and Analyzing knowledge intensive business processes with KMDL*. GITO mbH Verlag: Berlin, 2012.

Gronau, N.; Müller, C.; Uslar, M.: *The KMDL Knowledge Management Approach: Integrating Knowledge Conversions and Business Process Modeling*. In: Karagiannis, D.; Reimer, U. (Eds.), Practical Aspects of Knowledge Management, Springer Berlin / Heidelberg, 2004, pp. 1-10.

Habermas, J.: *Erkenntnis und Interesse: Mit einem neuen Nachwort*. Suhrkamp: Frankfurt/M., 1973.

Hackel, M.: *Auf dem Weg zum interdisziplinären mechatronischen Konstruktionsprozess. Entwickelnde Arbeitsforschung im Maschinen- und Anlagenbau*. Fernuniversität Hagen, Fakultät für Kultur- und Sozialwissenschaften, PhD Thesis. 2010

Hahn, A.; Gruening, F.; Hausmann, K.: Ontology-based product data management. In: *PDT Europe 2005 Proceedings* (2005).

Hansen, M.T.; Nohria, N.; Tierney, T.: What's your strategy for managing knowledge. In: *Harvard Business Review* 77 (1999), pp. 106-116.

Harashima, F.; Tomizuka, M.; Fukuda, T.: Mechatronics - "What Is It, Why, and How?" An Editorial. In: *IEEE/ASME Transactions on Mechatronics* 1 (1996), pp. 1-4.

Heath, T.; Bizer, C.: Linked data: Evolving the web into a global data space. In: *Synthesis lectures on the semantic web: theory and technology* 1 (2011), pp. 1-136.

Hehenberger, P.; Egyed, A.; Zeman, K.: *Hierarchische Designmodelle im Systementwurf mechatronischer Produkte.* In *Proceedings of the VDI Mechtronik*, Heidelberg, 2009.

Hellenbrand, D.; Lindemann, U.: *A Framework for Integrated Process Modeling and Planning of Mechatronic Products.* In *Proceedings of the International Conference on Engineering Design - ICED 11*, Copenhagen, 2011.

Henrichs, M.R.: *A Conceptual Framework for Constructing Distributed Object Libraries using Gellish.* Delft University of Technology, Faculty of Electrical Engineering, Mathematics and Computer Science, Master. 2009

Hewit, J.R.: *Mechatronics -The contributions of advanced control.* In *Proceedings of the Conference on Mechatronics and Robotics*, Düsseldorf, 1993.

Hitzler, P.; Krotzsch, M.; Rudolph, S.: *Foundations of semantic web technologies.* Chapman and Hall/CRC: Boca Raton, 2011.

Holsapple, C.W.; Joshi, K.D.: *A knowledge management ontology.* In: Holsapple, C.W. (Ed.), Handbook on knowledge management: Knowledge matters, Berlin Heidelberg, 2003, pp. 89-124.

Holsapple, C.W.; Joshi, K.D.: A formal knowledge management ontology: Conduct, activities, resources, and influences. In: *Journal of the American Society for Information Science and Technology* 55 (2004), pp. 593-612.

Holyoak, K.J.; Morrison, R.G.: *Thinking and Resoning: A Reader's Guide.* In: Holyoak, K.J.; Morrison, R.G. (Eds.), The Oxford handbook of thinking and reasoning, Oxford University Press, New York, 2012.

Hommes, B.-J.: *The Evaluation of Business Process Modeling Techniques.* TU Delft, 2004

Horváth, I.: A treatise on order in engineering design research. In: *Research in Engineering Design* 15 (2004), pp. 155-181.

Housel, T.J.; Bell, A.H.: *Measuring and managing knowledge.* McGraw-Hill Higher Education: New York, 2001.

Hubka, V.; Eder, W.E.: *Einführung in die Konstruktionswissenschaft: Übersicht, Modell, Anleitungen.* Springer: Berlin Heidelberg, 1992.

Hull, D.L.: *Science and Selection: Essays on Biological Evolution and the Philosophy of Science.* Cambridge University Press, 2001.

Isermann, R.: *Mechatronische Systeme.* Springer: Berlin Heidelberg, 2008.

ISYPROM Consortium, ISYPROM - Modellbasierte Prozess- und Systemgestaltung für die Innovationsbeschleunigung. Überblick über die Ergebnisse des Verbundprojektes, 2011a.

ISYPROM Consortium, ISYPROM Referenzprozess zur Systementwicklung, ISYPROM - Modellbasierte Prozess- und Systemgestaltung für die Innovationsbeschleunigung., 2011b.

ITQ GmbH: *ITQ Corporate Brochure: Competence in Mechatronics.* Retrieved 29.10.2010, from http://www.itq.de/files/itq_corporate_brochure.pdf.

Jansen, S.: *Eine Methodik zur modellbasierten Partitionierung mechatronischer Systeme.* Ruhr-Universität Bochum, Fakultät für Maschinenbau, PhD Thesis. 2006

Kahneman, D.: A perspective on judgment and choice: mapping bounded rationality. In: *American psychologist* 58 (2003), p. 697.

Kahneman, D.: *Thinking, fast and slow.* Penguin Books: London, 2011.

Kaiser, J.-M.; Conrad, J.; Köhler, C.; Wanke, S.; Weber, C.: *Classification of tools and methods for knowledge management in product development.* In *Proceedings of the International Conference on Engineering Design - ICED 08*, Dubrovnik, 2008.

Kebede, G.: Knowledge management: An information science perspective. In: *International Journal of Information Management* 30 (2010), pp. 416-424.

Khosrow-Pour, M., Dictionary of information science and technology, Idea Group Reference, Hershey, PA, 2007.

Kiefer, C.; Bernstein, A.: *Application and evaluation of inductive reasoning methods for the semantic Web and software analysis.* Reasoning Web. Semantic Technologies for the Web of Data, Springer, 2011, pp. 460-503.

Kim, S.; Hwang, H.; Suh, E.: A process-based approach to knowledge-flow analysis: a case study of a manufacturing firm. In: *Knowledge and Process Management* 10 (2003), pp. 260-276.

Klabunde, S.: *Wissensmanagement in der integrierten Produkt- und Prozessgestaltung.* Universität des Saarlandes, PhD Thesis. 2002

Klein, J.T.: *Interdisciplinarity: History, theory, and practice.* Wayne State Univ Press: Detroit, 1990.

Knorr Cetina, K.: *Epistemic Cultures. How the Sciences Make Knowledge.* Harvard University Press: Cambridge, MA, 1999.

Krause, F.-L.; Franke, H.-J.; Gausemeier, J.: *Innovationspotenziale in der Produktentwicklung.* Hanser Verlag: München Wien, 2007.

Kühnl, C.-P.: Smarte Maschinen braucht das Land. In: *Mechatronik* (2007), pp. 58-59.

Kultusministerkonferenz, Handreichung für die Erarbeitung von Rahmenlehrplänen der Kultusministerkonferenz für den berufsbezogenen Unterricht in der Berufsschule und ihre Abstimmung mit Ausbildungsordnungen des Bundes für anerkannte Ausbildungsberufe, Sekretariat der Kultusministerkonferenz Referat Berufliche Bildung und Weiterbildung, Bonn, 2007.

Kümmel, M.A.: *Integration von Methoden und Werkzeugen zur Entwicklung von mechatronischen Systemen.* Universität-Gesamthochschule Paderborn, PhD Thesis. 1999

Kusunoki, K.; Nonaka, I.; Nagata, A.: Organizational Capabilities in Product Development of Japanese Firms: A Conceptual Framework and Empirical Findings. In: *Organization Science* (1998), pp. 699-718.

Larses, O.; Adamsson, N.: *Drivers for model based development of mechatronic systems.* In *Proceedings of the International Conference on Engineering Design - ICED 04,* Dubrovnik, 2004.

Levy, M.: WEB 2.0 implications on knowledge management. In: *Journal of knowledge management* 13 (2009), pp. 120-134.

Liebowitz, J.; Rubenstein-Montano, B.; McCaw, D.; Buchwalter, J.; Browning, C.; Newman, B.; Rebeck, K.: The knowledge audit. In: *Knowledge and Process Management* 7 (2000), pp. 3-10.

Lossack, R.-S.: *Wissenschaftstheoretische Grundlagen für die rechnerunterstützte Konstruktion.* Springer: Berlin Heidelberg, 2006.

Love, T.: Philosophy of design: a meta-theoretical structure for design theory. In: *Design Studies* 21 (2000), pp. 293-313.

Lu, S.C.Y.; Liu, A.: Abductive reasoning for design synthesis. In: *CIRP Annals - Manufacturing Technology* 61 (2012), pp. 143-146.

Lückel, J.; Koch, T.; Schmitz, J.: *Mechatronik als integrative Basis für innovative Produkte.* VDI-Tagung: Mechatronik - Mechanisch/Elektrische Antriebstechnik, Wiesloch, 2000.

McGuinness, D.L.: *Ontologies Come of Age.* In: Fensel, D. (Ed.), Spinning the semantic web: bringing the World Wide Web to its full potential, MIT Press, Cambridge, MA, 2005, pp. 171-194.

McMahon, C.; Lowe, A.; Culley, S.: Knowledge management in engineering design: personalization and codification. In: *Journal of Engineering Design* 15 (2004), pp. 307-325.

Mendling, J.; Neumann, G.; van der Aalst, W.: *Understanding the Occurrence of Errors in Process Models Based on Metrics.* In: Meersman, R.; Tari, Z. (Eds.), Proceedings 15th International Conference on Cooperative Information Systems (CoopIS 2007), Springer Berlin / Heidelberg, 2007, pp. 113-130.

Mendling, J.; Reijers, H.A.; van der Aalst, W.M.P.: Seven process modeling guidelines (7pmg). In: *Information and Software Technology* 52 (2010), pp. 127-136.

Millbank, J. Mecha-what. *Mechatronics Forum Newsletter.* 1993, vol. 6. Available from.

Möhringer, S.; Stetter, R.: *A Research Framework for Mechatronic Design.* In Proceedings of the 11th International Design Conference DESIGN 2010, Dubrovnik – Croatia, 2010, pp. 885-894.

Nakamori, Y.: *Knowledge Science: Modeling the Knowledge Creation Process.* CRC Press, Taylor and Francis Group: Boca Raton, 2011.

Nattermann, R.S.; Anderl, R.: *Simulation Data Management Approach for Developing Adaptronic Systems – The W-Model Methodology.* In *Proceedings of the WASET 2011 International Conference in Software and Data Engineering (ICSDE)*, Bangkok, 2011, pp. 429-435.

Nemeth, C.P.: *Human Factors Methods for Design: Making Systems Human-Centered.* Taylor & Francis: Boca Raton, 2004.

Neubert, S.: Model Construction in MIKE (Model Based and Incremental Knowledge Engineering). In: *Knowledge Acquisition for Knowledge-Based Systems* (1993), pp. 200-219.

Neumann, F.: *Mechatronic Product Development: Potentials, Challenges, Terminology.* In *Proceedings of the 9th Workshop on Integrated Product Development*, Magdeburg/Germany, 2012.

Newman, B.B.: *Agents, artifacts, and transformations: The foundations of knowledge flows.* In: Holsapple, C.W. (Ed.), Handbook on knowledge management: Knowledge matters, Springer, Berlin Heidelberg, 2003, pp. 301-316.

Newman, B.B.; Conrad, K.; Carter, A.: *Knowledge Flow Modeling and Analysis with Focus on Enabling Actions and Decisions within the Business Process.* 44th Hawaiian International Conference on System Sciences, 2010.

Nickols, F.: *The Tacit and Explicit Nature of Knowledge: The Knowledge in Knowledge Management.* In: Cortada, J.W.; Woods, J.A. (Eds.), The knowledge management yearbook 2000-2001, Butterworth-Heinemann, Woburn, MA, 2000, pp. 12-21.

Nissen, M.E.: *Harnessing knowledge dynamics: Principled organizational knowing & learning.* Idea Group Inc.: Hershey, PA, 2006.

Nobre, A.L.: *Knowledge Processes and Organizational Learning.* In: McInerney, C.; Day, R. (Eds.), Rethinking Knowledge Management, Springer Berlin Heidelberg, 2007, pp. 275-299.

Nonaka, I.; Konno, N.: The Concept of "Ba": Building a Foundation for Knowledge Creation. In: *California Management Review* 40 (1998), pp. 40-54.

Nonaka, I.; Takeuchi, H.: *The knowledge-creating company: how Japanese companies create the dynamics of innovation.* Oxford University Press: New York, 1995.

Nonaka, I.; Toyama, R.; Konno, N.: SECI, Ba and Leadership: a Unified Model of Dynamic Knowledge Creation. In: *Long Range Planning* 33 (2000), pp. 5-34.

Nonaka, I.; von Krogh, G.: Perspective—Tacit knowledge and knowledge conversion: Controversy and advancement in organizational knowledge creation theory. In: *Organization Science* 20 (2009), pp. 635-652.

Nonaka, I.; von Krogh, G.; Voelpel, S.: Organizational Knowledge Creation Theory: Evolutionary Paths and Future Advances. In: *Organization Studies* 27 (2006), pp. 1179-1208.

OECD: *OECD Science, Technology and Industry Scoreboard.* OECD Paris, 2007.

OECD: *OECD Economic Surveys: Germany 2010.* OECD Paris, 2010.

Ovtcharova, J.; Weber, C.; Vajna, S.; Müller, U.: Neue Perspektiven für die Feature-basierte Modellierung. In: *VDI-Z* 139 (1997), pp. 34-37.

Oxford Dictionaries, Oxford Dictionaries, Oxford University Press, 2010.

Pahl, G.; Wallace, K.: *Engineering design: A systematic approach.* Springer: London, 2007.

Peirce, C.S. *Collected papers of Charles Sanders Peirce* [online]. InteLex, 1994. Available from.

Pogorzelska, B., Working Paper (detailed description) - KMDL® v2.2, A semi-formal description language for modelling knowledge conversions, University of Potsdam, Chair of Business Information Systems and Electronic Government, Potsdam, 2009.

Polanyi, M.: *The Tacit Dimension.* Doubleday & Company: Garden City, New York, 1966.

Pop, I.G.; Mătieş, V.: *Considerations about the mechatronical transdisciplinary knowledge paradigm.* In *Proceedings of the IEEE International Conference on Mechatronics*, Malaga, Spain, 2009 IEEE, pp. 387-390.

Pop, I.G.; Mătieş, V.: *Transdisciplinary Approach of the Mechatronics in the Knowledge Based Society.* Advances in Mechatronics, InTech Open Access Publisher, 2011, pp. 271-300.

Porter, M.E.: The Competitive Advantage of Nations. In: *Harvard Business Review* 2 (1990).

Porter, M.E.: *Competitive strategy: techniques for analyzing industries and competitors.* Free Press: New York, 1998.

Presley, A.; Huff, B.; Liles, D.H.: *A comprehensive enterprise model for small manufacturers.* In *Proceedings of the 2nd Industrial Engineering Research Conference*, Los Angeles, 1993, pp. 664-668.

Probst, G.J.B.; Raub, S.; Romhardt, K.: *Wissen managen: wie Unternehmen ihre wertvollste Ressource optimal nutzen.* Gabler: Wiesbaden, 2010.

Probst, G.J.B.; Romhardt, K.: *Bausteine des Wissensmanagements - ein praxisorientierter Ansatz.* In: Dr. Wieselhuber and partner (Ed.), Handbuch Lernende Organisation: Unternehmens- und Mitarbeiterpotentiale erfolgreich erschliessen, Gabler, Wiesbaden, 1997, pp. 129-144.

Qamar, A.; Wikander, J.; During, C.: *Designing Mechatronic Systems: A Model-Integration Approach.* In *Proceedings of the International Conference on Engineering Design - ICED 11*, Copenhagen, 2011.

Reich, Y.: The study of design research methodology. In: *Journal of Mechanical Design, Transactions Of the ASME* 117 (1995), pp. 211-214.

Reif, K.: *Automobilelektronik: eine Einführung für Ingenieure.* Vieweg und Teubner: Wiesbaden, 2009.

Repko, A.F.: *Interdisciplinary research: Process and theory.* SAGE Publications: Los Angeles, 2008.

Rijgersberg, H.; van Assem, M.; Top, J.: Ontology of units of measure and related concepts. In: *Semantic Web* 4 (2013), pp. 3-13.

Romhardt, K.: *Die Organisation aus der Wissensperspektive: Möglichkeiten und Grenzen der Intervention.* Université de Genève, PhD Thesis. 1998

Roozenburg, N.; Eekels, J.: *Product design: fundamentals and methods.* Wiley: Chichester, 1995.

Roth, D.; Binz, H.; Watty, R.: *Generic structure of knowledge within the product development process.* In *Proceedings of the International Conference on Engineering Design - ICED 10*, Dubrovnik, 2010, pp. 1681-1690.

Roth, K.: *Konstruieren mit Konstruktionskatalogen.* Springer: Berlin Heidelberg, 1994.

Roth, S., Innovationsstrategien erfolgreicher Automobilzulieferer, Materialien zur Automobilindustrie, VDA, 2008.

Russell, S.J.; Norvig, P.: *Artificial intelligence: a modern approach (3rd Edition)* Prentice Hall Englewood Cliffs, NJ, 2009.

Schauer, H.; Schauer, C.: *Modeling Techniques for Knowledge Management.* Knowledge Management Strategies: A Handbook of Applied Technologies, IGI, New York, 2008, pp. 91-115.

Schmidt, D.C.: Model-driven engineering. In: *IEEE computer* 39 (2006), pp. 25-31.

Schön, D.A.: Problems, frames and perspectives on designing. In: *Design Studies* 5 (1984), pp. 132-136.

School of Knowledge Science: *What is knowledge science?* Retrieved 29.02.2012, from http://www.jaist.ac.jp/ks/en/aboutus.html#knowledge.

Schreiber, G.: *Knowledge Engineering and Management: The CommonKADS Methodology.* MIT Press: Cambridge, MA, 2000.

Schurz, G., Die Bedeutung des abduktiven Schließens in Erkenntnis- und Wissenschaftstheorie, in: Schurz, G.; Hieke, A. (Eds.), IPS-Preprints, Department of Philosophy, University of Salzburg, Salzburg, 1995.

Schurz, G.: Patterns of abduction. In: *Synthese* 164 (2008), pp. 201-234.

Schweitzer, G.: *Mechatronik – Aufgaben und Lösungen: VDI-Berichte Nr. 787.* VDI-Verlag, Düsseldorf, 1989.

Sendler, U.: *Das PLM-Kompendium.* Springer: Berlin Heidelberg, 2009.

Siemens PLM: *Mechatronics Concept Designer.* Retrieved 19.01.2011, from http://www.plm.automation.siemens.com/en_us/Images/Siemens-PLM-NX-Mechatronics-Concept-Designer-fs_tcm1023-109227.pdf.

Siemens PLM, Mechatronics Concept Designer 7.5, Plano, TX, 2011.

Sinha, R.; Paredis, C.J.J.; Liang, V.-C.; Khosla, P.K.: Modeling and Simulation Methods for Design of Engineering Systems. In: *Journal of computing and information Science in Engineering* 1 (2001), pp. 84-91.

Smaili, A.; Mrad, F.: *Applied Mechatronics.* Oxford University Press: New York, 2008.

Snowden, D.: Organic Knowledge Management: Part I The ASHEN Model: an enabler of action. . In: *Knowledge Management* 3 (2000), pp. 14-17.

Society of Automotive Engineers: In: *SAE Transactions* (1940), p. 460.

Song, P.; Krovi, V.; Kumar, V.; Mahoney, R.: *Design and virtual prototyping of humanworn manipulation devices.* In *Proceedings of the ASME Design Engineering Technical Conferences,* Las Vegas, Nevada, 1999.

Staat, W.: On abduction, deduction, induction and the categories. In: *Transactions of the Charles S. Peirce Society* 29 (1993), pp. 225-237.

Stanovich, K.E.: *Distinguishing the reflective, algorithmic, and autonomous minds: Is it time for a tri-process theory.* In: Evans, J.S.B. (Ed.), In two minds: Dual processes and beyond, Oxford University Press, New York, 2009, pp. 55-88.

Stanovich, K.E.: *Rationality and the Reflective Mind.* Oxford University Press: New York, 2011.

Stanovich, K.E.; West, R.F.: Individual differences in reasoning: Implications for the rationality debate? In: *Behavioral and brain sciences* 23 (2000), pp. 645-665.

Stanovich, K.E.; West, R.F.; Toplak, M.E.: The complexity of developmental predictions from dual process models. In: *Developmental Review* 31 (2011), pp. 103-118.

Stark, R.; Damerau, T.: *Produktentstehung im Umbruch - notwendige Weichenstellungen für die Zukunft!* Innovationsforum Integrierte Systementwicklung - ISYPROM Abschlussveranstaltung, Berlin, 2011.

Steck, R.: EPLAN Engineering Center - Standardisieren durch Strukturieren. In: *CAD CAM* (2008), pp. 36-38.

Stetter, R.; Seemüller, H.; Chami, M.; Voos, H.: *Interdisciplinary System Model for Agent-supported Mechatronic Design.* In *Proceedings of the International Conference on Engineering Design - ICED 11*, Copenhagen, 2011.

Steup, M., The Analysis of Knowledge, in: Zalta, E.N. (Ed.), The Stanford Encyclopedia of Philosophy, The Metaphysics Research Lab, Stanford University, Stanford, 2008.

Steup, M., Epistemology, in: Zalta, E.N. (Ed.), The Stanford Encyclopedia of Philosophy, The Metaphysics Research Lab, Stanford University, Stanford, 2011.

Stokes, M.: *Managing engineering knowledge: MOKA: methodology for knowledge based engineering applications.* Professional Engineering Publishing: London, 2001.

Stork, A.: *FunctionalDMU: Towards experiencing behavior of mechatronic systems in DMU.* ProSTEP iViP symposium, Berlin, 2010.

Stuckenschmidt, H.: *Ontologien: Konzepte, Technologien und Anwendungen.* Springer: Berlin Heidelberg, 2009.

Sveiby, K.E.: *Frequently asked questions.* Retrieved 16.07.2001, from http://www.sveiby.com.au/faq.html.

Synek, P.-M.: Mechatronik – Chancen und Herausforderung für den Maschinenbau. In: *Konstruktion-Zeitschrift fur Konstruktion Entwicklung im Maschinen Apparate Geratebau* (2003), p. 40.

Szykman, S.; Racz, J.; Bochenek, C.; Sriram, R.D.: A web-based system for design artifact modeling. In: *Design Studies* 21 (2000), pp. 145-165.

Takeda, H.; Veerkamp, P.; Yoshikawa, H.: Modeling design processes. In: *AI magazine* 11 (1990), p. 37.

Tomiyama, T.; Takeda, H.; Yoshioka, M.; Shimomura, Y.: *Abduction for Creative Design*. In *Proceedings of the 15th Int. Conf. on Design Theory and Methodology–DTM'03, Proc. 2003 ASME Design Eng. Tech. Conf. & Comp. and Info. in Eng. Conf*, 2003, pp. 2-6.

Tomiyama, T.; Yoshioka, M.; Tsumaya, A.: *A Knowledge Operation Model of Synthesis*. In: Chakrabarti, A. (Ed.), Engineering Design Synthesis: Understanding, Approaches and Tools, Springer, London, 2002.

Tomizuka, M.: Mechatronics: from the 20th to 21st century. In: *Control Engineering Practice* 10 (2002), pp. 877-886.

Tomkinson, D.; Horne, J.: *Mechatronics engineering*. McGraw Hill: New York, 1996.

Torry-Smith, J.M.; Mortensen, N.H.: *A Mechatronic Case Study Highlighting the Need for Re-thinking the Design Approach*. In *Proceedings of the International Conference on Engineering Design - ICED 11*, Copenhagen, 2011.

Ullman, D.G.: *The Mechanical Design Process*. McGraw-Hill: New York, 2003.

Ushakov, D. Direct Modeling - Who and Why Needs It? A Review of Competitive Technologies. 2011. Available from: http://isicad.net/articles.php?article_num=14805.

Vajna, S. Basics for Process Navigation. 2006. Available from: http://www.pronavtec.de/inc/get.file.php?8.

Vajna, S.; Podehl, G.: Durchgängige Produktmodellierung mit Features. In: *CAD-CAM Report* 3 (1998), pp. 1–8.

Vajna, S.; Weber, C.; Bley, H.; Zeman, K.: *CAx für Ingenieure: eine praxisbezogene Einführung*. Springer: Berlin Heidelberg, 2009.

van Renssen, A.: *Gellish: a generic extensible ontological language - design and application of a universal data structure*. TU Delft, Electrical Engineering, Mathematics, Computer Science, PhD Thesis. 2005

VDA, Auto Jahresbericht 2005, Verband der Automobilindustrie, Frankfurt/M., 2005.

VDI: *VDI 2221: Systematic approach to the development and design of technical systems and products*. Beuth Verlag: Berlin, 1993.

VDI: *VDI 2206: Design methodology for mechatronic systems*. Beuth Verlag: Berlin, 2004.

VDI: *VDI 5610: Knowledge management for engineering*. Beuth Verlag: Berlin, 2009.

Venselaar, K.; van der Hoop, W.G.; van Drunen, P.: *The knowledge base of the designer*. In: Simons, P.R.J.; Beukhof, G. (Eds.), Regulation of Learning, SVO, 1987.

Visser, W.: *The Cognitive Artifacts of Designing*. L. Erlbaum Associates: Hillsdale, NJ, 2006a.

Visser, W.: Designing as construction of representations: A dynamic viewpoint in cognitive design research. In: *Human–Computer Interaction* 21 (2006b), pp. 103-152.

von Krogh, G.; Nonaka, I.; Rechsteiner, L.: Leadership in Organizational Knowledge Creation: A Review and Framework. In: *Journal of Management Studies* 49 (2012), pp. 240-277.

Wang, G.G.: Definition and review of virtual prototyping. In: *Journal of Computing and Information Science in Engineering (Transactions of the ASME)* 2 (2002), pp. 232-236.

Weinberger, D.: *Too Big to Know: Rethinking Knowledge Now That the Facts Aren't the Facts, Experts Are Everywhere, and the Smartest Person in the Room Is the Room*. Basic Books: New York, 2012.

Weinert, F.E.: *Concepts of competence*. In: OECD (Ed.), OCED-Project Definition and Section of Competencies: Theoretical and Conceptual Foundations (DeSeCo). Bundesamt für Statistik, Neuchâtel, 1999.

Welp, E.G.; Labenda, P.; Bludau, C.H.: *Usage of ontologies and software agents for knowledge based design of mechatronics systems*. In *Proceedings of the International Conference on Engineering Design - ICED 07*, Paris, 2007.

Whitman, L.E.; Huff, B.L.; Presley, A.R.: *The needs and issues associated with representing and integrating multiple views of the enterprise*. In *Proceedings of the IFIP TC5 WG5.3/5.7 Third International Working Conference on the Design of Information Infrastructure Systems for Manufacturing*, Fort Worth, Texas, 1999, pp. 139-152.

Wikander, J.; Törngren, M.; Hanson, M.: Mechatronics Engineering-Science and Education. In: *IEEE Robotics and Automation Magazine* 8 (2001).

Wilson, T.D.: The nonsense of knowledge management. In: *Information research* 8 (2002), pp. 8-1.

Wolf, F.; Rauhut, C.; Happ, S.; Buschow, C.; Dräger, K.: *Wissensmanagement im Enterprise 2.0 – Der Wikipedia Irrtum.* from http://www.besser20.de/prasentation-wissensmanagement-im-enterprise-20-der-wikipedia-irrtum-jetzt-online/75/.

Yares, E.: *CAD Usability and Model Deconstruction.* In: Insights on Engineering Tools [online]. 2011. Available from: http://www.evanyares.com/cad-usability-and-model-deconstruction/.

Zeigler, B.P.; Praehofer, H.; Kim, T.G.: *Theory of modeling and simulation.* Academic press New York, NY, 2000.

Zhuge, H.: *The knowledge grid.* World Scientific Publishing: Singapore, 2004.

Zins, C.: Conceptions of information science. In: *Journal of the American Society for Information Science and Technology* 58 (2007a), pp. 335-350.

Zins, C.: Conceptual approaches for defining data, information, and knowledge. In: *Journal of the American Society for Information Science and Technology* 58 (2007b), pp. 479-493.

Zirkler, S.C.: *Transdisziplinäres Zielkostenmanagement komplexer mechatronischer Produkte.* Technische Universität München, Lehrstuhl für Produktentwicklung, PhD Thesis. 2010

Zohm, F.: *Management von Diskontinuitäten. Das Beispiel der Mechatronik in der Automobilindustrie.* RWTH Aachen, Lehrstuhl für Produktionssystematik, PhD Thesis. 2003

Glossary

Agent: In *Artificial Intelligence*, an agent (also: intelligent agent) is an autonomous entity, which perceives its environment through its sensors and acts upon the environment through its actuators (Russell and Norvig 2009).

Component Structure[50]: The component structure represents the concrete technical artifact or system comprising the components, assemblies and their interconnections. It takes into consideration the spatial interrelationships and manufacturing related requirements (Pahl and Wallace 2007).

Continuous System: A continuous system (also: analog system or *Differential Equation System Specifications* (DESS)) shows a behavior that change continuously over time. Its state variables are determined by continuous functions of time (Khosrow-Pour 2007).

Discrete Event System: Within a discrete event system (also: *Discrete Event System Specifications* (DEVS)) the state variables change only at discrete points of time due to sudden events that are not previously know (Vajna, Weber, et al. 2009).

Discrete Time System: A discrete time system (also: digital system or *Discrete Time System Specifications* (DTSS)) shows a behavior observable through its state variables changing only at discrete points of time (Khosrow-Pour 2007).

Explicit Knowledge: Explicit knowledge refers to a form of knowledge that can be articulated through words, diagrams, formulae, computer programs, and similar means and can be readily transmitted to other people. Explicit knowledge can either be represented in the form of mental representations in the human brain or in its physical form by means of language or writing.

Function Structure: The function structure consists of (a) the subfunctions comprising the overall function of the mechatronic system, and (b) the flows of information, energy, and material linking these subfunctions.

Individual Knowledge Creation: Individual knowledge creation comprises all kinds of cognitive activities (e.g. reasoning, creative thinking, learning) leading to an extension of an individual's knowledge. New knowledge is created through interactions between individual and environment and through new combinations of existing knowledge.

Interdisciplinarity: Interdisciplinarity is a disciplinary interaction model characterized by the proactive interaction and integration of multiple disciplines working on a common topic. Through the interdisciplinary interactions, the issues and questions common to the involved disciplines will be linked and the existing disciplinary approaches will be restructured by explicit focusing and blending disciplines (Frodeman, Klein, et al. 2010).

---

[50] In German: „Baustruktur"

Interdiscipline: An interdiscipline designates a scientific field that starts in-between the bodies of knowledge of established disciplines and may later on become an academic discipline in its own right (Repko 2008). In addition to Mechatronics, other prominent interdisciplines include e.g. biochemistry, bioinformatics, biophysics, and systems engineering.

Knowledge Flow: A knowledge flow refers to the transfer of knowledge between different participants as sender and receiver each affiliated with an organizational level (individual, group, organization). A knowledge flow may transform the representation of knowledge along the explicitness dimension. An associated KM activity (create, organize, formalize, share, apply, refine) further characterizes each knowledge flow taking place. (Cf. four-dimensional knowledge flow model (Nissen 2006))

Knowledge Management: Knowledge management comprises a set of strategies, methods, technologies, and tools for the purposeful and methodical management of knowledge creation, storage, sharing, and its application (Deng 2007; Kebede 2010). It aims at realizing the full potential of knowledge in all kinds of knowledge-intensive business processes (e.g. decision making, problem solving) at all organizational levels (personal, group, organization, and inter-organization) (Kebede 2010).

Mechatronic Product Development: Mechatronic product development comprises the integrated engineering activities targeting the development of mechatronic systems. It involves an interdisciplinary collaboration typically between mechanical engineering, electrical engineering, and computer science with control engineering as one of its sub-disciplines. It applies a systems oriented design approach, i.e. it considers the objects of design activities as technical systems and adopts procedures, methods, and tools of systems engineering accordingly. Additionally, it follows the principles of model-based development, i.e. it broadly uses models, model-based tools, and analysis techniques to support its different activities. (Neumann 2012)

Mechatronic System: A mechatronic system is a technical system that consists of functionally and/or spatially integrated mechanical, electronic and software components. The integration of these heterogeneous components is achieved first by a multidisciplinary and later by an interdisciplinary collaboration of the relevant engineering disciplines (typically: mechanical engineering, electrical engineering, and computer science with control engineering as one of its sub-disciplines) that enable synergistic effects resulting in new and upgraded features as well as innovative solutions.

Model-based Development: *"Model-based development refers to a development approach whose activities emphasise the use of models, tools and analysis techniques for the documentation, communication and analysis of decisions taken at each stage of the development lifecycle."* (El-khoury 2006)

Multidisciplinarity: Multidisciplinarity is a disciplinary interaction model described as a *"juxtaposition of disciplines"* in a strictly additive manner without any explicit cooperation between the different disciplines. The relationships between the disciplines remain limited and temporary, and no exchange of scientific methods and procedures will take place (Klein 1990). The different disciplinary results will simply be added, but not merged to an integrated result (Zirkler 2010).

Organizational knowledge creation: *"Organizational knowledge creation is the process of making available and amplifying knowledge created by individuals, as well as crystallizing and connecting it with an organization's knowledge system."* (Nonaka, von Krogh, et al. 2006)

Partitioning (also: **Technology Allocation** or **Domain Allocation**): Partitioning is an activity allocating the overall system functions to different domains and to different units and modules (VDI 2004). This activity can be separated in (a) functional partioning that allocates one or several functions to one or multiple domains, and (b) spatial partioning that consists of the geometrical positioning and spatial grouping of components within the mechatronic system (Jansen 2006).

Polymorphic (also: **multi-faceted) Concept**: A classical definition, which is universally valid in all possible contexts, cannot sufficiently classify polymorphic concepts. Depending on the context of interpretation, these types of concepts typically have several definitions. The concepts of information and knowledge have a polymorphic nature (Aamodt and Nygård 1995).

Requirements Traceability: *"refers to the ability to describe and follow the life of a requirement, in both a forwards and backwards direction (i.e., from its origins, through its development and specification, to its subsequent deployment and use, and through periods of on-going refinement and iteration in any of these phases)."* (Gotel and Finkelstein 1994)

Solution Concept[51]: The solution concept comprises the overall physical and logical operating characteristics and the types and structure of the components (VDI 2004).

Synergy: Interaction of two or more elements creating an effect that is larger than the summation of their single contributions (Oxford Dictionaries 2010).

Tacit Knowledge: Tacit knowledge refers to a form of knowledge that is difficult to articulate and formalize through means like language or writing, because it is bound both to an individual and to a particular context (Nonaka and Takeuchi 1995).

Virtual Prototyping: *"By virtual prototyping, we refer to the process of simulating the user, the product, and their combined (physical) interaction in software through the different stages of product design, and the quantitative performance analysis of the product."* (Song, Krovi, et al. 1999; cited in Wang 2002)

Wirk-Structure[52]: The Wirk-structure consists of the identified Wirk-principles[53] and solution elements for each of the subfunctions forming the function structure. Usually, this mapping is not trivial and the Wirk-structure will therefore have a significantly different topology to the function structure.

---

[51] In German: „Lösungskonzept"

[52] In German: „Wirkstruktur"

[53] In German: „Wirkprinzipien"

List of Appendices

Appendix 1: Analysis of proposed process elements in MPD

Appendix 2: Analysis of proposed roles in MPD

Appendix 3: Process modeling of MPD (KMDL Process View) for selected representative knowledge-intensive process elements

Appendix 4: Modeling of knowledge-intensive tasks within MPD (KMDL Activity View)

Appendix 5: Logical system architecture of design support system – detailed views

Process Modules / Process Elements	Bellalouna (2006), Integration platform	Follmer, Hehenberger, et al. (2011), Model-based mechatronic design	Gausemeier and Kahl (2010), Design methodology for self-optimizing mechatronic systems	ISYPROM (2011b), Reference process	Jansen (2006), Partitioning	VDI 2206 (2004)
System Design — Derive functions from requirements		• Determine functions		•		• Abstraction for identifying the main problems
Establish function structure			• Draw up function hierarchy	•		•
Conceptual design at system level (functional partitioning): Search of Wirk-principles and solution elements for each subfunction resulting in Wirk-structure	(•) Establish logical system architecture	• Search for principle solutions and their combination	• Identify "solution patterns" (Wirk-principle and patterns in computer science) or alternatively, reuse existing solution elements for each subfunction.	• For each combination of Wirk-principles and solution elements	•	•
Spatial partitioning into subsystems and components resulting in component structure	(•) Establish technical system architecture	• Divide into realizable modules	(•) Based on the Wirk-structure, an initial component structure can be developed		•	• Concretizing of Wirk-structure resulting in component structure
Conceptual design at subsystem level			Identify "solution patterns" (Wirk-principle and patterns in computer science) or alternatively, reuse existing solution elements for each subfunction.	• Component-based design of subsystems		•
Analyze and assess variants for conceptual design of subsystems				•		
Establish and verify one or multiple variants of the system design concept consisting of: - Function structure - Wirk-principles/solution elements - Wirk-structure - Preliminary component structure - Assessed solution variants of subsystems				•		
Select one of the variants of the system design concept				•		

Process Modules	Process Elements	Bellalouna (2006), Integration platform	Follmer, Hehenberger, et al. (2011), Model-based mechatronic design	Gausemeier and Kahl (2010), Design methodology for self-optimizing mechatronic systems	ISYPROM (2011b), Reference process	Jansen (2006), Partitioning	VDI 2206 (2004)
System Design	Derive domain-spanning system design consisting of: - Function structure - Wirk-principles/solution elements - Wirk-structure - Component structure - Definition of interfaces			(•) Integration of the modules' principle solutions into a detailed principle solution of the whole system. Determine contradictions between the principle solutions of the modules. Technical-economical evaluation of the solution.	(•)		•
Domain-Specific Design	Derive domain-specific function structures from system-level function structure (implicit partitioning)					•	
Domain-Specific Design	Establish E/E system architecture	• Establish E/E system architecture		(•) Establish E/E specific models			
Domain-Specific Design	Establish software system architecture	• Establish hardware system architecture •		(•) Establish control engineering specific models (•) Establish software engineering specific models			
Domain-Specific Design	Establish mechanical system architecture	•		(•) Establish mechanical engineering specific models	•		
System Integration	Stepwise integration of components and subsystems to overall system						•
System Integration	Conduct integration tests for subsystems and the integrated system at different levels: - Functional testing, testing for interferences, regression tests, performance testing - Interface tests - Tests of spatial integration						

Process Modules	Process Elements	Bellalouna (2006), Integration platform	Follmer, Hehenberger, et al. (2011), Model-based mechatronic design	Gausemeier and Kahl (2010), Design methodology for self-optimizing mechatronic systems	ISYPROM (2011b), Reference process	Jansen (2006), Partitioning	VDI 2206 (2004)
System Integration	Identify incompatibilities between subsystems						•
	Eliminate incompatibilities between subsystems						•
	Determine optimal overall solution according to requirements						•
	Establish software baseline	•					
	Establish E/E baseline	*Establish E/E baseline*					
	Establish hardware baseline	*Establish hardware baseline*					
	Establish mechanical configuration	•					
	Establish configuration of integrated system	•					•
Overall Activities	Modeling and model analysis		*Create mechatronic system model either using top-down or bottom-up modeling* •				
	Assurance of properties according to requirements		•				•
	Version interdisciplinary system	•					
	Release interdisciplinary system	•					

Roles	Bellalouna (2006), Integration platform	Follmer, Hehenberger, et al. (2011), Model-based mechatronic design	Gausemeier and Kahl (2010), Design methodology for self-optimizing mechatronic systems	ISYPROM (2011b), Reference process	Jansen (2006), Partitioning	VDI 2206 (2004)
Requirements manager						
System architect				•		
Product/system manager				•		
Head of function				•		
System analyst				•		
Mechanical design engineer	•			(•) Component developer		
E/E system architect						
E/E design engineer	•			(•) Component developer		
Software architect						
Software developer	•			(•) Component developer		
Test engineer				•		
System integrator				•		
Process designer				•		

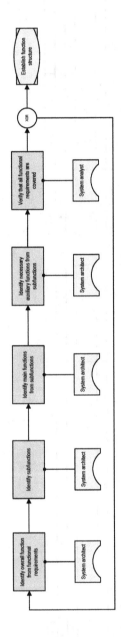

KMDL Process View for "Derive functions from requirements"

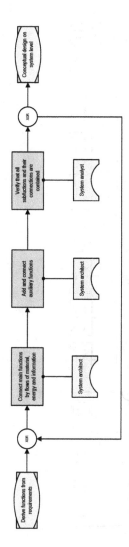

KMDL Process View for "Establish function structure"

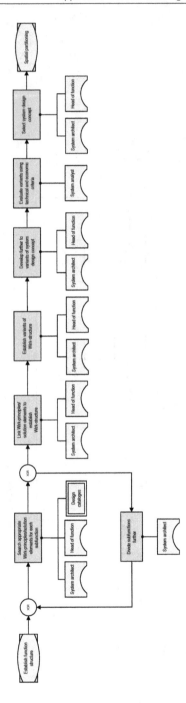

KMDL Process View for "Conceptual design at system level"

In order to keep the complexity of the process views at a manageable level, only the case of implicit partitioning will be considered.
During implicit partitioning the association to a domain is implicitly established by assigning a Wirk-principle/solution element to each subfunction (Jansen 2006).

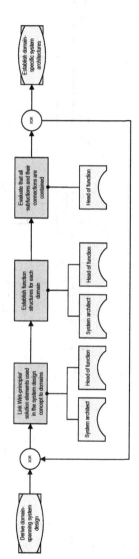

KMDL Process View for "Derive domain-specific function structures from system-level function structure"

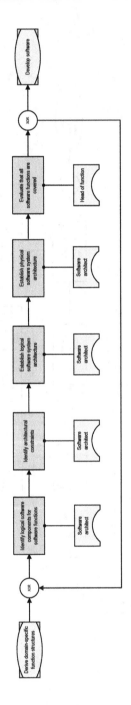

KMDL Process View for "Establish software system architecture"

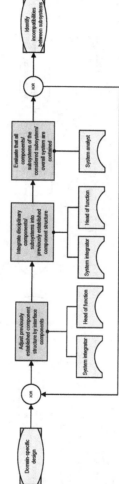

KMDL Process View for "Stepwise integration of components and subsystems"

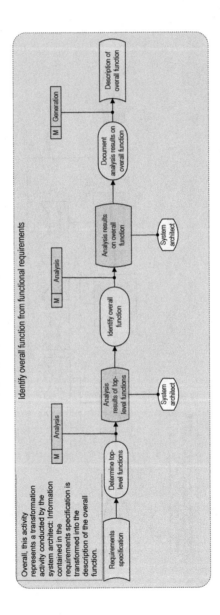

KMDL Activity View for Process Element "Derive functions from requirements"

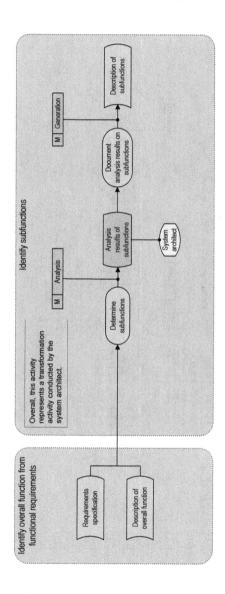

KMDL Activity View for Process Element "Derive functions from requirements"

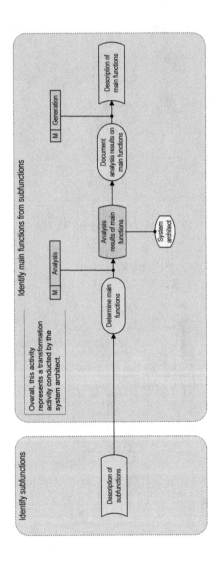

KMDL Activity View for Process Element "Derive functions from requirements"

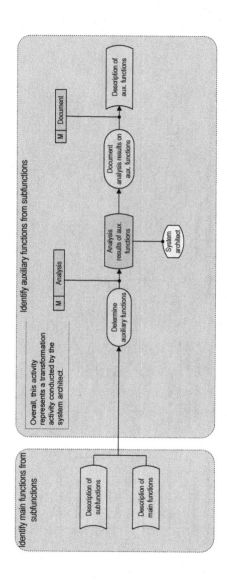

KMDL Activity View for Process Element "Derive functions from requirements"

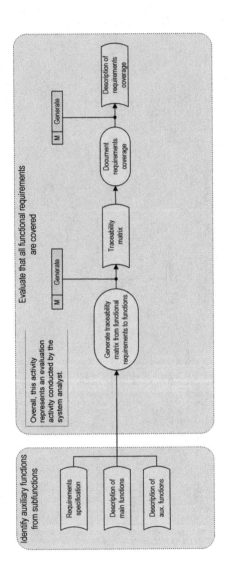

KMDL Activity View for Process Element "Derive functions from requirements"

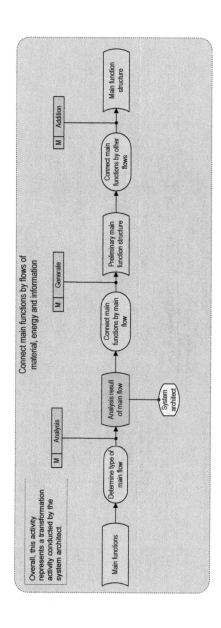

KMDL Activity View for Process Element "Establish function structure"

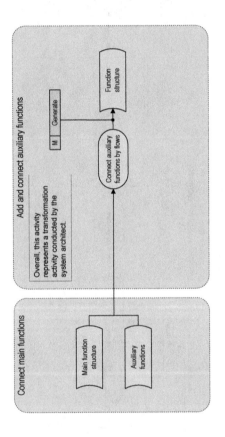

KMDL Activity View for Process Element "Establish function structure"

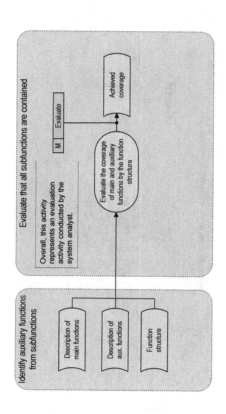

KMDL Activity View for Process Element "Establish function structure"

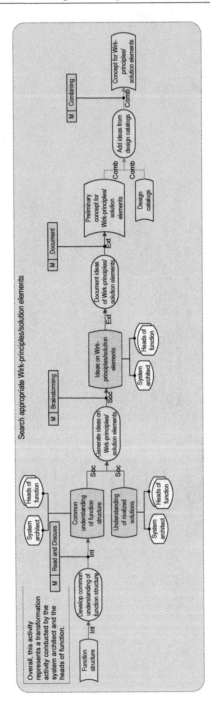

KMDL Activity View for Process Element "Conceptual design at system level"

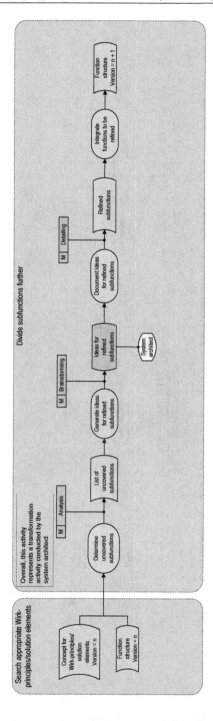

KMDL Activity View for Process Element "Conceptual design at system level"

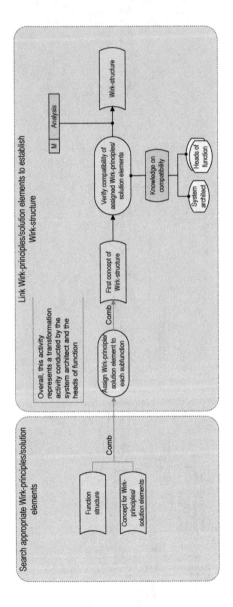

KMDL Activity View for Process Element "Conceptual design at system level"

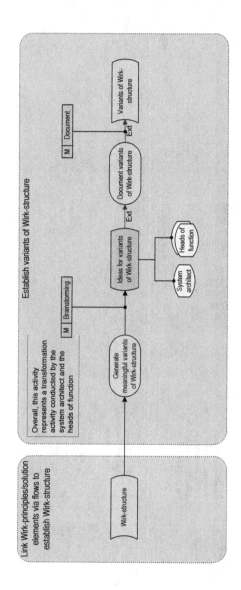

KMDL Activity View for Process Element "Conceptual design at system level"

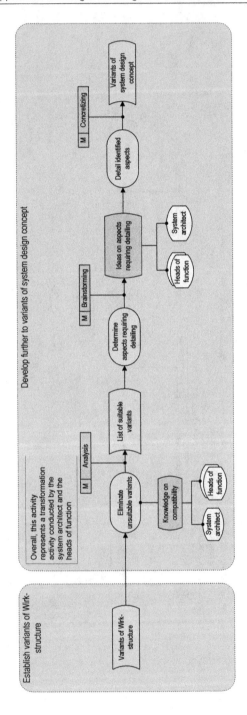

KMDL Activity View for Process Element "Conceptual design at system level"

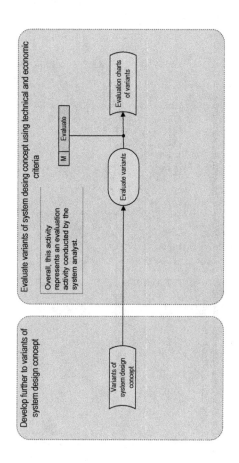

KMDL Activity View for Process Element "Conceptual design at system level"

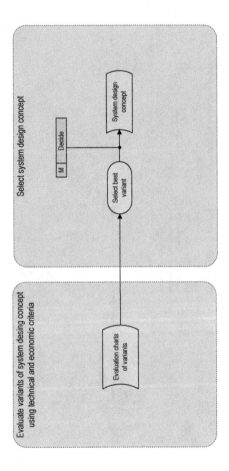

KMDL Activity View for Process Element "Conceptual design at system level"

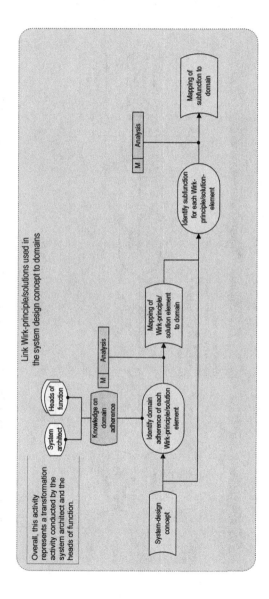

KMDL Activity View for Process Element "Derive domain-specific function structures from system-level function structure"

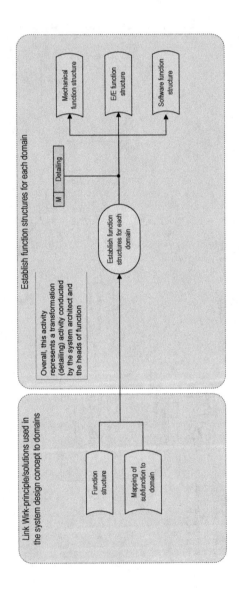

KMDL Activity View for Process Element "Derive domain-specific function structures from system-level function structure"

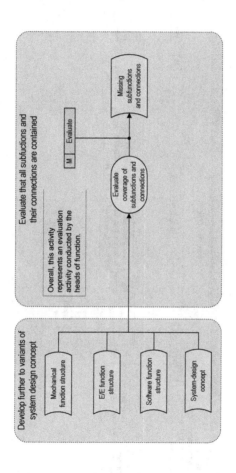

KMDL Activity View for Process Element "Derive domain-specific function structures from system-level function structure"

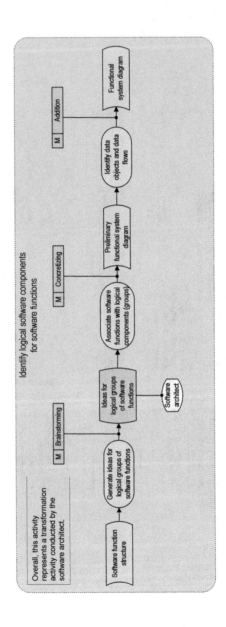

KMDL Activity View for Process Element "Derive domain-specific function structures from system-level function structure"

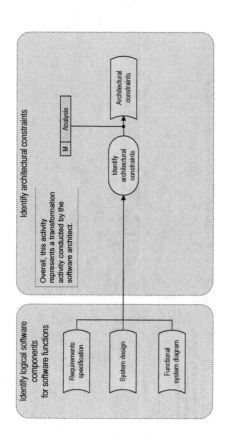

KMDL Activity View for Process Element "Derive domain-specific function structures from system-level function structure"

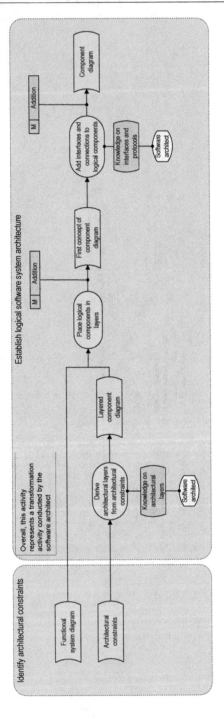

KMDL Activity View for Process Element "Establish software system architecture"

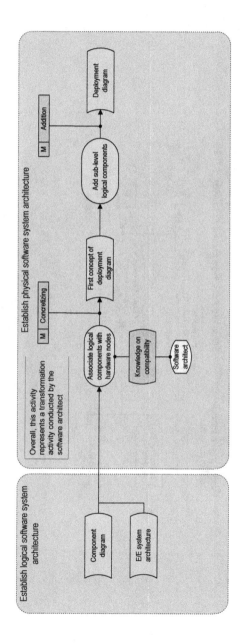

KMDL Activity View for Process Element "Establish software system architecture"

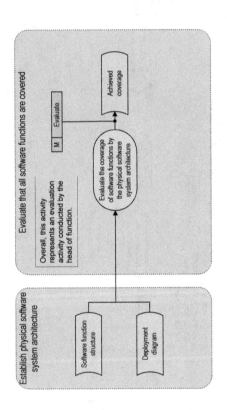

KMDL Activity View for Process Element "Establish software system architecture"

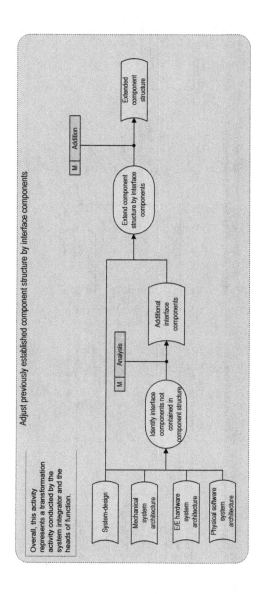

KMDL Activity View for Process Element "Stepwise integration of components and subsystems to overall system"

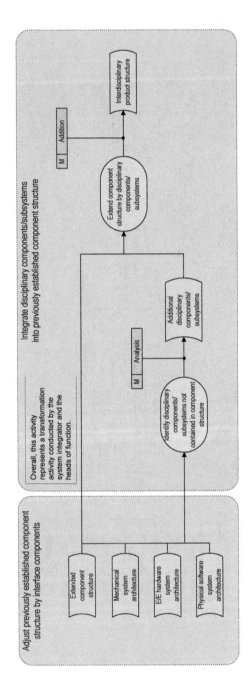

KMDL Activity View for Process Element "Stepwise integration of components and subsystems to overall system"

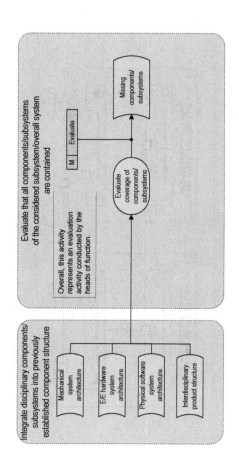

KMDL Activity View for Process Element "Stepwise integration of components and subsystems to overall system"

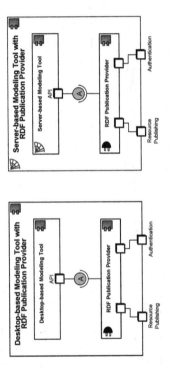

Logical System Architecture for "Modeling Tools"

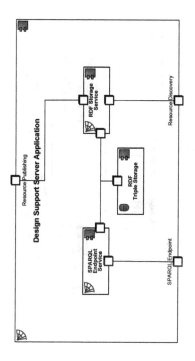

Logical System Architecture for "Design Support Server"

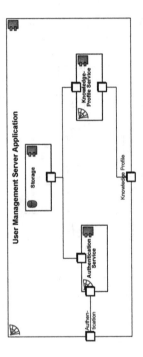

Logical System Architecture for "User Management Server"

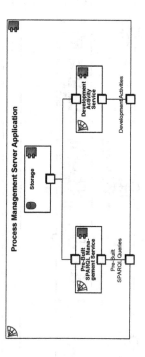

Logical System Architecture for "Process Management Server"

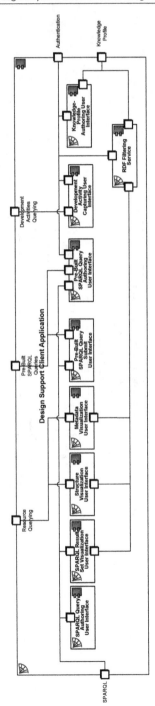

Logical System Architecture for "Design Support Client"

Printed in the United States
By Bookmasters